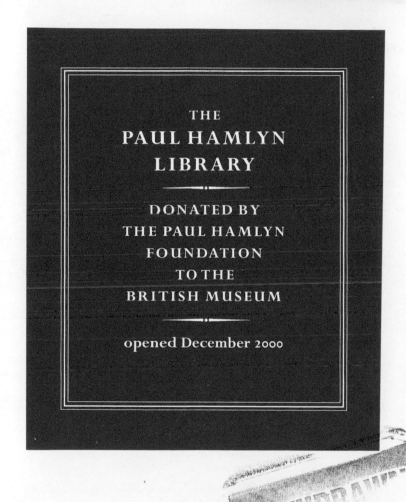

Contrary to prevailing opinion, the roots of modern science were planted in the ancient and medieval worlds long before the Scientific Revolution of the seventeenth century. Indeed, that revolution would have been inconceivable without the cumulative antecedent efforts of three great civilizations: Greek, Islamic, and Latin. With the scientific riches it derived by translation from Greco-Islamic sources in the twelfth and thirteenth centuries, the Christian Latin civilization of Western Europe began the last leg of the intellectual journey that culminated in a scientific revolution that transformed the world.

Four essential factors enabled medieval Europe to prepare the way for the new science of the seventeenth century: translations into Latin of Greek and Arabic scientific texts in the twelfth and thirteenth centuries; the development of universities, which were uniquely Western and used the translations as the basis of a science curriculum; the adjustments of Christianity to secular learning; and the transformation of Aristotle's natural philosophy. This study reviews the accomplishments of medieval science and also carefully considers how they looked forward to the Scientific Revolution.

# THE FOUNDATIONS OF MODERN SCIENCE
## IN THE MIDDLE AGES

CAMBRIDGE HISTORY OF SCIENCE

Editors

GEORGE BASALLA
*University of Delaware*

OWEN HANNAWAY
*Johns Hopkins University*

Physical Science in the Middle Ages
EDWARD GRANT

Man and Nature in the Renaissance
ALLEN G. DEBUS

The Construction of Modern Science: Mechanisms and Mechanics
RICHARD S. WESTFALL

Science and the Enlightenment
THOMAS L. HANKINS

Biology in the Nineteenth Century: Problems of Form,
Function, and Transformation
WILLIAM COLEMAN

Energy, Force, and Matter: The Conceptual Development of
Nineteenth-Century Physics
P. M. HARMAN

Life Science in the Twentieth Century
GARLAND E. ALLEN

The Evolution of Technology
GEORGE BASALLA

Science and Religion: Some Historical Perspectives
JOHN HEDLEY BROOKE

Science in Russia and the Soviet Union: A Short History
LOREN R. GRAHAM

Science and the Practice of Medicine in the Nineteenth Century
W. F. BYNUM

# THE FOUNDATIONS OF MODERN SCIENCE IN THE MIDDLE AGES

*Their religious, institutional, and intellectual contexts*

EDWARD GRANT

*Indiana University*

CAMBRIDGE
UNIVERSITY PRESS

PUBLISHED BY THE PRESS SYNDICATE OF THE UNIVERSITY OF CAMBRIDGE
The Pitt Building, Trumpington Street, Cambridge CB2 1RP, United Kingdom

CAMBRIDGE UNIVERSITY PRESS
The Edinburgh Building, Cambridge CB2 2RU, United Kingdom
40 West 20th Street, New York, NY 10011-4211, USA
10 Stamford Road, Oakleigh, Melbourne 3166, Australia

First published 1996

Printed in the United States of America

Typeset in Palatino

*Library of Congress Cataloging-in-Publication Data*
Grant, Edward, 1926–
The foundations of modern science in the Middle Ages / Edward
Grant.
p.   cm. – (Cambridge history of science)
Includes bibliographical references and index.
ISBN 0-521-56137-X. – ISBN 0-521-56762-9 (pbk.)
1. Science, Medieval.   I. Title.   II. Series.
Q124.97.G68 1996
509'.4'0902 – dc20                                    95–26530
                                                       CIP

*A catalog record for this book is available from
the British Library*

ISBN 0 521 56137 X hardback
ISBN 0 521 56762 9 paperback

To
my colleagues
past and present
in the
Department of History and Philosophy of Science
Indiana University-Bloomington

# Contents

x                                 *Contents*

# *Preface*

Twenty-five years have passed since the publication of *Physical Science in the Middle Ages* in 1971 in the John Wiley History of Science Series, and nineteen years since Cambridge University Press assumed responsibility for the series in 1977. In the early 1980s, I was asked to revise the book, but other duties and responsibilities made the task unfeasible. When I finally had the opportunity a few years ago, the prospect of doing a revision no longer seemed inviting. Too much had happened in the interim. Merely expanding the old version with new material – and there has been much of it since 1971 – while retaining the earlier structure and general outlook was, frankly, unappealing. My sense of the medieval achievement in science and natural philosophy and my understanding of the intellectual environment that produced it, as well as my perception of the relationship between medieval science and the Scientific Revolution, had all been fundamentally transformed.

Between 1902 and 1916, Pierre Duhem, a famous French physicist turned historian, wrote fifteen volumes on medieval science. Duhem was the first to blow away the dust of centuries from manuscript codices that had lain untouched since the Middle Ages. What he discovered led him to make the startling claim that the Scientific Revolution, associated with the glorious names of Nicholas Copernicus, Galileo Galilei, Johannes Kepler, René Descartes, and Isaac Newton, was but an extension and elaboration of physical and cosmological ideas formulated in the fourteenth century, primarily by Parisian masters at the University of Paris. Duhem regarded medieval scholastic natural philosophers as Galileo's precursors. By his numerous publications, Duhem made medieval science a respectable research field and placed the late Middle Ages in the mainstream of scientific development. He thus filled the hiatus that had existed between Greek and Arabic science, on the one extreme, and early modern science in seventeenth-century Europe, on the other. For the first time, the history of science was provided with a genuine sense of continuity.

Duhem's claims seemed extravagant to most historians of science, and even to many medievalists, who often suffered from an inferiority complex about the expression "medieval science," which, until the appearance of Pierre Duhem, was regarded by many scholars as an oxymoron. Medievalists bold enough to pursue Duhem's approach, or to try other ways, also had to contend with the charge of "Whiggism," namely, that they were selecting from the mass of medieval science and natural philosophy ideas that sounded modern and that could be proclaimed as anticipations of things to come. Medieval claims were further subverted by Alexandre Koyré, a preeminent historian of the Scientific Revolution, who insisted that the classical science of the seventeenth century was in no way a continuation of medieval physics, even when medieval ideas and concepts were strikingly similar to ideas proposed in the Scientific Revolution. It was, he argued, a "decisive mutation" (*mutation décisive*).[1] The ideas and concepts were embedded in radically different intellectual contexts. Or, to use language made famous by Thomas Kuhn in *The Structure of Scientific Revolutions*, the respective *normal paradigms* of medieval and seventeenth-century physics were *incommensurable*. The physics and cosmology of the Middle Ages, it was argued, were based wholly upon Aristotelian natural philosophy, which was incompatible with the new science that emerged in the seventeenth century. Indeed, Aristotelian natural philosophy was viewed as the major obstacle to the birth of the new science. Only by its repudiation could the Scientific Revolution have succeeded.

*Physical Science in the Middle Ages* was written with the conviction that this interpretation was essentially correct, and that the Middle Ages had not contributed significantly to the Scientific Revolution of the seventeenth century. True, there were some medieval anticipations of later achievements, especially in problems concerning motion. But these were insufficient to prompt me, and most historians of medieval science, to attribute any meaningful role to medieval science and natural philosophy in bringing about the new science of the seventeenth century.

A few years ago, it occurred to me that perhaps we – historians of medieval science and of the Scientific Revolution – had interpreted the medieval contribution too narrowly. We had judged the Middle Ages in terms of the specific influences it might have exerted on this or that science, usually physics, and on whether it had played a role in reshaping scientific methodology. But the significance of these alleged influences, or contributions, was often called into question, although a number of them will be cited in this study. Because it was difficult to demonstrate that these medieval "anticipatory" ideas had actually exerted any direct influence, most scholars found little reason to assume a medieval contribution. The case seemed closed.

My attitude changed dramatically, however, when, some years ago, I asked myself whether a Scientific Revolution could have occurred in the seventeenth century if the level of science in Western Europe had remained what it was in the first half of the twelfth century. That is, could a scientific revolution have occurred in the seventeenth century if the massive translations of Greco-Arabic science and natural philosophy into Latin had never taken place? The response seemed obvious: no, it could not.[2] Without the translations, many centuries would have been required before Western Europe could have reached the level of Greco-Arabic science, thus delaying any possibility of a transformation of science. But the translations did occur and so did the Scientific Revolution. It follows that something happened between approximately 1200 and 1600 that proved conducive for the production of a scientific revolution. If these medieval facilitations did not occur in the exact science themselves, then they must be identified elsewhere.

Although scholars are still at work determining whether specific medieval discussions and achievements in physics (here I am thinking especially of concepts of motion and matter theory) and cosmology contributed to the new science of the seventeenth century, some of which will be mentioned later, the claims in this study do not depend on such specific influences. Even if the Middle Ages made few significant contributions to the advancement of the sciences themselves, or none at all, the claims made here should stand on their own. But if no noteworthy medieval contributions were made to help shape specific scientific advances in the seventeenth century, in what ways did the Middle Ages contribute to the Scientific Revolution and, more to the point, lay the foundations for it? Whatever these contributions may have been, they had to be of a permanent nature and also had to be new to Western Europe, in the sense that they did not exist prior to the twelfth century. What these foundational elements were will be described and elaborated in chapter 8.

In light of my radically altered perception of the significance of medieval science and natural philosophy, nothing less than a new book seemed appropriate, the burden of which is to validate the new claims. Nevertheless, I have included some material from the earlier work, especially material from chapter 4 ("The Physics of Motion"). But the approach in this book is radically different. Indeed, this volume might even be viewed as supplementing the earlier book. Where *Physical Science in the Middle Ages* sought to convey the essential and substantive features and contributions of medieval science, while acquiescing in the old interpretation, the present study, though not challenging the old, narrow interpretation, expands the discussion to a much broader canvas. It views the substantive achievements of the Middle Ages within a broad societal and institutional setting that includes the translations, Christianity, and the universities. What emerges is a new interpretation that attributes a major role to the Middle Ages in the generation of early modern science, a role that is independent of whether or not medieval scholars made identifiable contributions to the transformation of the exact sciences in the Scientific Revolution.

I intend this study to be a relatively brief, interpretive essay for a broad audience. Therefore, I have limited footnotes almost exclusively to the identification of quotations. Also, I would like to thank the anonymous reader who reviewed my manuscript for the Cambridge University Press. Despite our radically different attitudes toward history in general, and the Middle Ages in particular, I profited from a number of helpful suggestions and corrections and am still awed by the diligence and dedication with which he or she carried out this difficult assignment.

# 1

## The Roman Empire and the first six centuries of Christianity

DURING the first four centuries of Christianity, the Roman Empire was a geographical colossus, extending from the Atlantic Ocean in the west to Persia in the east, and from Britain in the north to regions south of the Mediterranean Sea. Within this Greco-Roman world, Christianity was born and disseminated. Its birth and early development occurred in a period of vast religious change and economic upheaval. For the first two hundred years of its existence, Christianity was no more visible and noticeable than many other of the numerous mystery religions and cults that had proven attractive to people at all levels of society. The sense of comfort that pagans derived from their belief in the traditional Homeric and Roman gods of the state religions was disappearing. The new cults – for example, Isis, Mithras, Cybele, and Sol Invictus (Unconquered Sun), as well as Gnosticism and Christianity – that were replacing the traditional deities not only borrowed ideas and rituals from one another but also came to share a few basic beliefs. The world was evil and would eventually pass away. Humans, sinful by nature, could achieve never-ending bliss only if they turned away from the things of this world and cultivated those of the eternal spiritual realm. Along with practicing varying degrees of asceticism, many of the cults believed in a redeemer god who would die in order to bring eternal life after death to his faithful followers. Contemporary philosophical schools, such as Neoplatonism and Neopythagoreanism, were also affected by these popular currents. Some came to function as religions, as they sought to guide their adherents toward salvation and union with God, even employing magic to achieve their ends. (The philosophical schools, however, were ill-suited to the competition, because they required lengthy periods of study and training before they judged a student capable of understanding the world and its governance.) The reaction to centuries of homogenized and impersonal worship of traditional gods took the form of a desire for a single, personal god, the ruler of the world, with whom one could establish an intimate, personal relationship. Many came to believe that they could be transformed by a direct revelation from the one god, a revelation that would enable them to overcome the

1

evils of the world. Numerous groups emerged, each concerned with its own private and exclusive program for salvation. Christianity was one of these.

How Christianity triumphed over the traditional gods and also over the numerous other mystery religions and cults that were its rivals cannot be discussed here. But certain features that concern Christianity's dissemination and its attitude toward the larger Roman world around it are germane to the subsequent development of science and therefore central to this volume. A notable feature of the spread of Christianity was the slowness of its dissemination. The spread of Christianity beyond the Holy Land and its surrounding region began in earnest after Saint Paul proselytized into the Gentile world, especially into Greece, during the middle of the first century. In retrospect – and by comparison with the spread of Islam – the pace of the dissemination of Christianity appears quite slow. Not until A.D. 300 was Christianity effectively represented throughout the Roman Empire. And not until 313, in the reign of Constantine, was the Edict of Milan (or Edict of Toleration) issued, which conferred on Christianity full legal equality with all other religions in the empire. In 392, the Emperor Theodosius not only ordered pagan temples closed but also proscribed pagan worship, which thereafter was classified as treason. Thus it was not until 392, or the end of the fourth century, that Christianity became the exclusive religion supported by the state. After almost four centuries, Christianity was triumphant. It had taken nearly four centuries (approximately three hundred fifty years from the beginning of Paul's serious efforts to spread Christianity wherever possible) to achieve this result. By contrast, Islam, following the death of Muhammad in 632, was carried over an enormous geographical area in a remarkably short time. In less than one hundred years, Islam was the dominant religion from the Arabian peninsula westward to the Straits of Gibraltar, northward to Spain, and eastward to Persia, Balkh, Bukhara, Samarkand, and Khwarizm. But where Islam was spread largely by conquest during its first one hundred years, Christianity spread slowly and, with the exception of certain periods of persecution, relatively peacefully. It was the slow percolation of Christianity that enabled it to adjust to the pagan world around it and thus prepare itself for a role that could not have been envisioned by its early members.

## CHRISTIANITY AND PAGAN LEARNING

The momentous adjustment of Christianity to the pagan world around it is manifested by numerous learned Christians whose writings were subsequently influential. To Gregory of Nyssa, Christianity was "the sublime philosophy." Yet he, like many other eminent Christians, recognized that pagan philosophy still had a role to play, as did pagan tradition and learning generally. In the process of acquiring an education, Christians came to share numerous cultural traditions with their pagan neighbors and fellow citizens. Much of this came by way of *paideia*, a kind of shared civility, which "offered ancient, almost proverbial guidance, drawn from the history and literature of Greece, on serious issues, issues which no notable – Christian or polytheist, bishop or layman – could afford to ignore: on courtesy, on the prudent administration of friendship, on the control of anger, on poise and persuasive skill when faced by official violence."[1]

Because they came from varied backgrounds, the Greek fathers, who significantly shaped Christian attitudes toward pagan philosophy, were hardly of one mind on the subject. Some were hostile to science and philosophy out of concern for their potentially subversive effect on the faith. Most, however, denounced these disciplines because of their conviction that Christianity was "the sublime philosophy" and therefore the only system capable of delivering truth. For many of them, science was a confusing, contradictory body of knowledge. Church fathers like Tatian, Eusebius, Theodoret, and Saint Basil seemingly delighted in subverting Greek science by showing that its conclusions were often fatuous or contradictory. Theodoret likened science to writing on water,[2] and Basil declared, perhaps in imitation of Plato's scornful description of the Presocratics, that "the wise men of the Greeks wrote many works about nature, but not one account among them remained unaltered and firmly established, for the later account always overthrew the preceding one. As a consequence there is no need for us to refute their words; they avail mutually for their own undoing."[3] Many church fathers, of whom Gregory of Nyssa was one, followed Plato and argued that science could at best give only probable knowledge, not genuine truth.

As early as the end of the second century and first half of the third century, other Christian apologists came to a quite different conclusion, arguing instead that Christianity could profitably utilize pagan Greek philosophy and learning. In a momentous move, Clement of Alexandria (ca. 150–ca. 215) and his disciple Origen of Alexandria (ca. 185–ca. 254) laid down the basic approach that others would follow. Greek philosophy was neither inherently good nor bad, but was one or the other depending on how it was used by Christians. Although the Greek poets and philosophers had not received direct revelation from God, they did receive natural reason and were thus heading toward truth. Philosophy – and secular learning in general – could thus be used to prepare the way for Christian wisdom, which was the fruit of revelation. Philosophy and science could be studied as "handmaidens to theology," that is, as aids in understanding Holy Scripture, an attitude that had already been advocated by Philo Judaeus, a resident of the Jewish community of Alexandria, early in the first century A.D. Science was thus regarded as a study that was preparatory for the higher disciplines that were concerned with Scripture and theology. In the second half of the fourth century, Basil of Caesarea reinforced the handmaiden idea in a brief treatise to students titled *On How to Make Good Use of the Study of Greek Literature*. Like many of his early Christian colleagues, however, Basil was ambivalent. He warned about the dangers of some of the great works of Greek literature, but he also recognized that a Christian could profit from familiarity with pagan writings and quoted from various Greek works. Much later, Christian humanists in the Renaissance viewed Basil's treatise as providing encouragement for Christians to study pagan Greek literature. Leonardo Bruni (1370–1444) was inspired to translate Basil's treatise into Latin because he saw in it justification for his own translations of Plutarch and Plato from Greek to Latin.

The handmaiden concept of Greek learning was widely adopted and became the standard Christian attitude toward secular learning. That Christians chose to accept pagan learning within limits was a momentous decision. They might have heeded Tertullian (ca. 150–ca. 225), who asked pointedly, "What indeed has Athens to do with Jerusalem? What concord is there between the Academy and the Church?" With the total triumph of Christianity at the end of the fourth century, the Church might have reacted against Greek pagan learning in general, and Greek philosophy in particular, finding much in the latter that was unacceptable or perhaps even offensive. They might have launched a major effort to suppress pagan learning as a danger to the Church and its doctrines. But they did not. Why not?

Perhaps it was in the slow dissemination of Christianity. After four centuries as members of a distinct religion, Christians had learned to live with Greek secular learning and to utilize it for their own benefit. Their education was heavily infiltrated by Latin and Greek pagan literature and philosophy. Numerous converts to Christianity – the most notable being Saint Augustine – had been steeped in pagan learning, which formed a normal part of their societal and cultural milieu. Although Christians found certain aspects of pagan culture and learning unacceptable, they did not view them as a cancer to be cut out of the Christian body.

The handmaiden theory was obviously a compromise between rejection of traditional pagan learning and its full acceptance. By approaching secular learning with caution, Christians could utilize Greek philosophy – especially metaphysics – to better understand and explicate Holy Scripture and to cope with the difficulties generated by the assumption of the doctrine of the Trinity and other esoteric dogmas. Ordinary daily life also required use of the mundane sciences such as astronomy and mathematics. Christians realized that they could not turn their backs on Greek learning. But many were also wary of pagan Greek science and philosophy, which contained ideas and concepts that were contrary to Christian doctrine. Among these ideas were the common Greek notion that the world was eternal and had no beginning and the deterministic interpretations of the world advocated by Stoic philosophers and astrologers, who often assumed a world rigidly determined by the configurations of the planets and stars. Like Saint Basil a few years before him, Saint Augustine (354–430), who was enormously influential in the Latin Middle Ages, reflects both attitudes. He advocated the study of the liberal arts, including the sciences – geometry, arithmetic, astronomy, and music – embodied in the traditional quadrivium of the seven liberal arts. But he was suspicious of astronomy, a discipline that frequently led its practitioners to astrological determinism, which he deplored. Augustine's ambivalence toward secular learning is reflected in his *Retractions*, written in 426, four years before his death, where he expressed regret that he had ever emphasized the study of the seven liberal arts and concluded that the theoretical sciences and mechanical arts are of no use to a Christian.

## HEXAEMERAL LITERATURE: CHRISTIAN COMMENTARIES ON THE CREATION ACCOUNT IN GENESIS

Although Christians adopted the handmaiden approach to science, science itself was not a major concern of theirs. However, their need to understand the Bible better and to explicate the creation account in Genesis made it advisable for Christians to learn something about natural philosophy and science. Following the pattern established by Philo Judaeus (d. ca. 40 A.D.), who left the first commentary on the creation account in Genesis, a number of influential church fathers – Saints Basil, Ambrose, and Augustine, for example – wrote commentaries that proved influential in the Middle Ages.

Basil (ca. 331–379), who wrote in Greek, presented his commentary in the form of nine homilies, delivered originally as lectures to audiences in a church. In this famous work, Basil sought to praise the glory and power of God and to instill in Christians a strong sense of moral purpose. To achieve these ends, he appealed to nature, as God's handiwork. In the process, he found it necessary to convey a modicum of contemporary scientific knowledge about the basic structure and composition of the world. For example, in explaining the words "In the beginning God created the heavens and the earth," Basil was compelled to consider a host of topics: whether creation was simultaneous, or in time; whether the heavens were created before the earth; the nature of the heavenly substance; the meaning of the firmament; the meaning of the waters above and below that firmament; clouds, vapors, and the four elements; the location and shape of the world; the production of vegetation on the earth; the creation of the planets and stars; and the creation of crawling creatures, birds, and sea life. Thus Basil confronted the question: how does the earth remain immobile at the center of the world? On what does it rest? Perhaps drawing on Aristotle, Basil considered a number of possible answers: the earth rests on air, or on water, or on something heavy. Rejecting these options – for example, if a heavier object supported the earth, one would then have to ask what held up the heavier object, and so on – Basil concluded that the earth has no reason to move because it lies in the middle of everything.[4] He never tired of emphasizing the marvelous design that God embedded in nature.

Basil frequently mixed his descriptions of natural phenomena and design, especially of the behavior of animals and plants, with morality. As he put it, "Everything in existence is the work of Providence, and nothing is bereft of the care owed to it. If you observe carefully the members even of the animals, you will find that the Creator has added nothing superfluous, and that He has not omitted anything necessary."[5] He drew lessons from the migration of fish, the stealth of the octopus, the function of the elephant's trunk, the behavior of dogs tracking wild animals, and the existence of both poisonous and edible plants. All play their designated role in nature, even poisonous plants, for, as Basil argued, "there is no one plant without worth, not one without use. Either it provides food for some animal, or has been sought out for us by the medical profession for the relief of certain diseases."[6] Thus did Basil respond to those who wondered why God would create poisonous plants capable of killing humans.

Basil's ideas were enormously influential in the western and eastern parts of the Roman Empire. In the West, Saint Ambrose (ca. 339–397), who possessed a sufficient knowledge of Greek to make use of Basil's homilies in his own Latin hexaemeral treatise, was instrumental in introducing Basil's ideas into the Latin language. (Basil's treatise was translated into Latin in the fifth century and was known directly in the Middle Ages.) It was Saint Augustine, however, who composed the most formidable and influential early Latin commentary on the creation account in Genesis. Not only was his much lengthier than those by Basil and Ambrose (he was familiar with Ambrose's), but it was also much more informative and philosophical. It had a considerable impact in the late Middle Ages, especially on the theological students who were required to write commentaries on Peter Lombard's *Sentences*, the second book of which was concerned with the creation and about which we shall say more later.

Basil also had an impact on Greek writers in the East, especially on John Philoponus, a Christian commentator on Aristotle of the sixth century. Philoponus's hexaemeral treatise was incomparably more sophisticated than Basil's. In defending the Mosaic account in Genesis against the traditional pagan Greek description of the physical world, Philoponus found it necessary to discuss numerous scientific claims and arguments. When his treatise became available in Western Europe in the sixteenth century, it made a significant impact.

Although these early Christian authors subordinated science and the study of nature to the needs of religion, they often indicated an interest in nature, as did Basil, that transcended the mere ancillary status that the study of nature was customarily accorded. The attitude of theologians toward natural philosophy during the late Middle Ages is eloquent testimony that invocation of the handmaiden theory eventually became little more than formulaic.

## CHRISTIANITY AND GRECO-ROMAN CULTURE

Greco-Roman culture and learning was sometimes viewed with suspicion, but it was not considered an enemy and its potential utility was recognized early on. Indeed, it may have received unintended support from the Christian attitude toward the state. Because they believed that the kingdom of heaven was imminent, early Christians paid relatively little heed to the world around them. They sought generally to meet their obligations to the state insofar as these did not violate their religious scruples. Nowhere is this better exemplified than in Jesus' reply to the Pharisees who sought to trap him by asking whether they should pay taxes to the Roman emperor. Their question presented Jesus with a dilemma: if he urged them not to pay taxes, he would be guilty of treason to the state; but if he urged them to pay, he would antagonize Jewish nationalists. Jesus' reply was momentous when he urged that they "Render therefore unto Caesar the things which are Caesar's; and unto God the things that are God's" (Matt. 22.21). Thus did Jesus acknowledge the state and implicitly urge his followers to be good citizens.

From the outset, Christians recognized the state as distinct from the church, although, as the Roman Catholic Church became more centralized, various popes sought to dominate the multiplicity of states in Europe. They based their arguments on the conviction that, by the nature of things, the priesthood had to be closer to God than did secular rulers. In a letter to Anastasius, the Eastern emperor, Pope Gelasius (492–496) declared that "there are ... two by whom principally this world is ruled: the sacred authority of the pontiffs and the royal power. Of these the importance of the priests is so much the greater, as even for Kings of men they will have to give an account in the divine judgment."[7] The later pretensions of the papacy were based on this notion. The pope claimed supremacy over emperors and kings, as when Innocent III (1198–1216) declared that "The Lord Jesus Christ has set up one ruler over all things as his universal vicar, and as all things in heaven, earth and hell bow the knee to Christ, so should all obey Christ's vicar, that there be one flock and one shepherd" and again when he proclaimed that "The *sacerdotium* [priestly power] is the sun, the *regnum* [royal power] is the moon. Kings rule over their respective kingdoms, but Peter rules over the whole earth. The *sacerdotium* came by divine creation, the *regnum* by man's cunning."[8]

A counterattack by the secular power, whether the Holy Roman Emperor or one or another of the kings of Europe, when it came, usually involved an invocation of Christ's statement that one ought to give to Caesar, or to the state, what is Caesar's, and give to God what is God's; or that Christ sat on David's throne as king and not on Aaron's throne as high priest; or that Christ would eventually rule the human race as king, not priest.[9]

From the fifth century through the late Middle Ages, the struggle for supremacy between the papacy and the numerous secular rulers with which it had to contend was ongoing. The power of the papacy reached its high point during the early thirteenth century with the pontificate of Innocent III, after which it declined, largely because secular rulers became so wealthy and powerful that they could no longer be controlled from Rome.

Significant here, however, is not which of these two contending powers was dominant at any time, but rather that each acknowledged the independence of the other. They regarded themselves as two swords, although, all too often, the swords were pointed at each other. Even when the Church asserted supremacy over the state, however, it never attempted to establish a theocracy by appointing bishops and priests as secular rulers. The tradition of the Roman state within which Christianity developed and the absence of explicit biblical support for a theocratic state were powerful constraints on unbridled and grandiose papal ambitions and, above all, made the imposition of a theocratic state unlikely. Although church and state were not as rigidly separated in the Middle Ages as they are today in the United States and Western Europe, and the two often interacted, even blatantly intervening in each other's affairs, they were, nonetheless, independent entities. Pope Gelasius's words cited earlier – "there are ... two by whom principally this world is ruled: the sacred authority of the pontiffs and the royal power" – bear witness to the separation.

Why are the relationships between early Christianity and Greek science and philosophy on the one hand, and between the Christian church and the secular state on the other hand, relevant to the history of science? Because, as we shall see, the separation of church and state, at least in principle, and, more significantly, the Christian accommodation with Greek science and philosophy, were instrumental conditions that facilitated the widespread, intensive study of natural philosophy during the late Middle Ages. As a consequence of the emergence of natural philosophy within the unique university system of the Latin Middle Ages, the revolutionary developments in science of the sixteenth and seventeenth centuries were made possible. We may better appreciate the force of these claims by a comparison of Western European developments with developments in two major contemporary civilizations, Islam and the Byzantine Empire. The differences are striking and will be described in the final chapter.

## THE STATE OF SCIENCE AND NATURAL PHILOSOPHY DURING THE FIRST SIX CENTURIES OF CHRISTIANITY

To comprehend the state of science that obtained by the beginning of the seventh century, it is essential to sketch the basic events that transformed the Roman Empire. During its first two centuries, from the reign of Caesar Augustus to the death of Marcus Aurelius, the Romans controlled a vast empire in which two languages were dominant. In the west, not surprisingly, the Romans succeeded in imposing a basic Roman culture in which Latin was the common means of communication, overlaying a multiplicity of native languages in the regions of Italy, Gaul, Spain, Britain, and North Africa. In the eastern part of the Roman Empire, which to a considerable extent coincided with the old Hellenistic world left in the wake of the conquests of Alexander the Great (that is, Greece, Asia Minor, Syria, Persia, Palestine, and Egypt), Greek was the common language. Beginning with the emperor Diocletian (284–305), the Roman Empire was split administratively into eastern and western parts, largely reflecting the linguistic split into Greek- and Latin-speaking regions. Diocletian chose a colleague, Maximianus, to rule in the West, while he governed in the East, establishing a new capital at Nicomedia. In 330, Constantine established yet another new capital in the East, Constantinople, locating it at the site of an old Greek colony, Byzantium, a name that would eventually stand for the empire itself. For a brief period, between 394 and 395, Theodosius the Great reunited the empire, ruling as sole emperor. Following his death in 395, however, the empire was again ruled by two independent, self-proclaimed emperors, one in the East and one in the West. The line of emperors in the West ended in 476, when Romulus Augustulus was deposed. But even with German states functioning in the Western empire after 476, the Roman Empire was still viewed as intact, and German rulers often acknowledged the empire by either taking or accepting the honored title of consul. This state of affairs continued until Charlemagne was crowned "Emperor of the Romans" by Pope Leo III on December 25, 800, thus beginning the long history of the Holy Roman Empire in Western Europe. By the time of Charlemagne's coronation, Western Europe had long ceased to be a de facto part of the Roman Empire. In the East, however, Roman emperors reigned continuously from the time of Constantine's foundation of Constantinople until the city fell to the Turks in 1453, more than one thousand years later. Thus did the Roman Empire fall for the final time.

Although Latin was the language of the Romans, and Roman military might had created a vast empire, the language of learning in the Roman Empire was Greek. In this sense, Athens conquered Rome. Latin-speaking Romans with intellectual pretensions, never a large group, usually learned Greek, and some went to Greece for their education.

How did science fare within the Roman Empire? Despite much political and military turmoil, the multiplication of mystery religions, and an emphasis on the occult, some of the greatest scientific works of the ancient world were written in this period (as always in the Greek language and in the eastern half of the empire). A few of these works exerted a profound influence on the later course of medieval science and well beyond into the Renaissance.

The first century A.D. saw the significant works of Hero of Alexandria (who wrote on pneumatics, mechanics, optics, and mathematics), Nicomachus (on Pythagorean arithmetic), and Theodosius and Menelaus (who both wrote on spherical geometry; Menelaus's *Spherics* is especially important for the treatment of spherical triangles and trigonometry). The greatest works in astronomy and medicine were written in the second century. Claudius Ptolemy wrote the *Mathematical Syntaxis*, or *Almagest*, as it was called by the Arabs, the greatest treatise in the history of astronomy until the time of Copernicus in the sixteenth century. Ptolemy's scientific genius was not confined to astronomy. He also wrote technical works in optics, geography, and stereographic projection, and he even produced the greatest of all astrological works, the *Tetrabiblos* (known in Latin as the *Quadripartitum*, the four-parted work). In the medical and biological sciences, Galen of Pergamum produced about one hundred fifty works embracing both theory and practice, which formed the basis of medical theory and study until the sixteenth and seventeenth centuries. Even in the third century significant contributions were made in mathematics by Diophantus in algebra and later by Pappus, who not only wrote commentaries on the great mathematical works of Greek antiquity but also, in his *Mathematical Collection*, showed originality and understanding of a high order.

The Greek world of late antiquity also contributed powerfully to natural philosophy, largely by way of commentaries on the works of Aristotle. Because Aristotelian natural philosophy plays a central role in this study, and because the Greek commentaries on Aristotle's works in late antiquity were of particular importance for the subsequent history of science, a brief description of the late Greek commentators will be given in the next chapter.

The achievements of the first six centuries of the Christian era were typical of the manner in which Greek science and natural philosophy had developed and advanced. Always the product of a small number of gifted scholars concentrated in a few centers, Greek science was a fragile enterprise, able to advance and preserve itself just so long as the intellectual environment was favorable, or at least not overtly antagonistic. Greek science at its traditional best in the Roman Empire was but a continuation of the progress already made in the physical and biological sciences of classical Greece and the Hellenistic world, when the works of Plato, Aristotle, Hippocrates, Eudoxus, Euclid, Archimedes, Appollonius of Perga, Hipparchus, Theophrastus, Herophilus, and Erastistratus established the highest levels of achievement.

As in our own day, however, there existed in antiquity an audience of educated individuals interested in the physical world but with little inclination or ability to tackle forbidding scientific treatises of a theoretical and abstract nature. To meet the needs of this group, scientific popularizers simplified and rendered palatable conclusions from the exact sciences and natural philosophy, which were then incorporated into handbooks and manuals. Greek authors began the process of popularization in the Hellenistic period. Not surprisingly, some of these treatises were filled with contradictory information. Readers who were astute enough to detect the inconsistencies were left to reconcile them as best they could.

Greeks who were instrumental in shaping the handbook tradition were the polymath Eratosthenes of Cyrene (ca. 275–194 B.C.), who supplied much geographical knowledge to the tradition; Crates of Mallos (fl. 160 B.C.); and especially Posidonius (ca. 135–51 B.C.), whose numerous works have not survived, but whose opinions on meteorology, geography, astronomy, and other sciences were absorbed into later handbooks to become permanent fixtures in the tradition. Continuing in the manner of Posidonius were other Greek authors, such as Geminus (ca. 70 B.C.); Cleomedes (first or second century A.D.), who wrote the astronomical and cosmological work *On the Cyclic Motions of the Celestial Bodies*; and Theon of Smyrna (first half of the second century A.D.), who wrote the *Manual of Mathematical Knowledge Useful for an Understanding of Plato* in which the whole universe is discussed, just as it is in Plato's *Timaeus*. Theon drew upon astronomy and cosmology as well as Pythagorean arithmetic and mathematics.

Commentaries on Plato's *Timaeus* constituted a significant part of the handbook tradition from the Hellenistic period to the early Middle Ages. Because the *Timaeus* was a scientific treatise concerned not only with the cosmos but also with the biological status of the human species, it was an admirable vehicle for the handbook tradition, and physical and biological themes could be appropriately included.

Following their conquest of Greece, Roman gentlemen were brought into contact with Greek culture during the second and first centuries B.C. By this time, the Greek handbook tradition was established and its treatises were admirably adapted to cater to Roman cultural interests. For although the Romans were awed by Greek intellectual accomplishments, they had little interest in theoretical and abstract science. When fashion dictated that cultured Romans become acquainted with the results of Greek science, the handbook method was there to meet the need. Romans who knew Greek consulted the Greek handbooks directly, but the great majority of Romans absorbed their knowledge through Latin translations or summaries. Soon, Latin authors began compiling their own handbooks on science.

Although the Latin encyclopedic tradition actually began in the first century B.C. with Marcus Terrentius Varro (116–27 B.C.), its two most significant early representatives were Seneca (d. A.D. 68) and Pliny the Elder (A.D. 23/24–79). In *Natural Questions*, Seneca concerned himself largely with geography and meteorological phenomena (for example, rainbows, halos, meteors, thunder, and lightning), after the manner of Aristotle's *Meteorology*. He drew heavily upon Aristotle, Posidonius, perhaps his major authority, Theophrastus, and other Greek sources. Because Seneca frequently drew morals from natural phenomena, his book was popular with Christians. He also transmitted to the Middle Ages an estimate of the size of the earth that was small enough to encourage men like Columbus and others to think that the oceans were sufficiently narrow to be readily navigable. Seneca also struck an optimistic note on the progress of science and knowledge when he predicted that continuous research would reveal nature's secrets.

Pliny's *Natural History* in thirty-seven books was a remarkable scissors-and-paste collection of enormous scope and detail. By his own estimate, he examined about two thousand volumes drawn from 100 authors. In Book I, Pliny presents a detailed outline of the topics and a full list of the authorities used for each of the thirty-six volumes that follow. Thus did he honor his predecessors. A total of 473 authors are listed, of whom presumably the 100 mentioned in his estimate were primary and some of the others were either known through intermediaries or perhaps used cursorily for isolated facts. Book II is devoted to cosmography; Books III to VI to regional geography; and Book VII to human generation, life, and death. Books VIII to XXXII are concerned with zoology and botany, including fabulous animals and the curative powers associated with animals and plants, and Books XXXIII to XXXVII consider mineralogy. As an indefatigable compiler, Pliny emphasized the curious and the odd in natural phenomena. Although confusions, inconsistencies, and misunderstandings abound in his work, the weakest sections are those that attempt to explain Greek theoretical science, which Pliny scarcely comprehended.

If Pliny's work was confused and frequently inconsistent, it was at least the product of great diligence coupled with an honest respect for the sources that provided the grist for his insatiable mill. Few of his successors shared his finer instincts. Although Pliny acknowledged many, if not most, of his sources, he would not have been thought immoral had he not done so. Plagiarism was not regarded as an intellectual crime. It would be inappropriate to apply our modern standards on plagiarism to the ancient and medieval worlds when incorporating passages from someone else's treatise was not considered reprehensible, nor was it censured by custom or practice. In the compilations of Pliny's late ancient and early medieval successors, passages and sections were often extracted from the works of others without acknowledgment. Thus Solinus, who lived in the third or fourth century A.D., compiled the encyclopedic *Collection of Remarkable Facts*, about which the most remarkable fact is that Solinus lifted most of it from Pliny. Solinus's treatise was so thoroughly raided in its turn that modern scholars are frequently unable to determine whether Pliny or Solinus was the source of this or that later opinion. Encyclopedic authors looked upon available handbooks as storehouses of information in the public domain that could be extracted, embellished, and rearranged to suit their purposes. The final products were then paraded as learned treatises drawn directly from the original sources. The scientific works and opinions of the likes of Plato, Aristotle, Archimedes, Euclid, Theophrastus, and others were cited repeatedly in the handbooks, as if the compilers had direct knowledge of them. It is all too apparent, however, that these encyclopedists had no direct acquaintance with the great scientific authors of the past and were but repeating – and frequently distorting – what earlier compilers had already repeated and distorted from their predecessors.

Between the fourth and eighth centuries, encyclopedic authors produced a series of Latin works that were to have significant influence throughout the Middle Ages, especially prior to 1200. Among this group, the most important were Chalcidius, Macrobius, Martianus Capella, Boethius, Cassiodorus, Isidore of Seville, and Venerable Bede. Chalcidius (fl. ca. fourth century A.D.) translated most of Plato's *Timaeus* into Latin and added a commentary whose astronomical portions he drew from Theon of Smyrna's *Manual*, mentioned earlier. Macrobius (fl. 400 A.D.), a Neoplatonist, incorporated encyclopedic learning into a commentary on Cicero's *Dream of Scipio*, which is actually Book VI of Cicero's *Republic*. Martianus Capella (fl. 410–439) wrote the popular *Marriage of Philology and Mercury*, an ornate, florid account of the seven liberal arts and a pale reflection of classical learning and wisdom.

Ancius Manlius Severinus Boethius (ca. 480–524) was one of the best of the Latin encyclopedists and also an unusual one because he knew Greek, although the extent of his knowledge is uncertain. Boethius wrote on the "quadrivium" (a term he may have introduced for the four mathematical sciences of the seven liberal arts), of which only his treatises on music and Pythagorean arithmetic survive, the latter in the form of a free translation of Nicomachus's *Introduction to Arithmetic*, originally composed in Greek. To these Boethius added his translations of some of Aristotle's logical treatises, perhaps Euclid's *Elements*, and unspecified works of Archimedes that have not survived. His commentaries on certain of the philosophical treatises that he translated and his most famous work, *On the Consolation of Philosophy*, written in prison while he awaited execution, were very influential. Cassiodorus (ca. 488–575) included sections on the seven liberal arts in his *Introduction to Divine and Human Readings* and was reasonably scrupulous about citing his authorities. Isidore of Seville (ca. 560–636), in addition to writing a treatise titled *On the Nature of Things*, compiled a vast encyclopedia called *The Etymologies*, in which he discussed the seven liberal arts, medicine, zoology, the mechanical arts, metallurgy, and other topics. Finally, there is Venerable Bede (ca. 673–735), perhaps the most intelligent of the Latin encyclopedists. In addition to a conventional encyclopedia, *On the Nature of Things*, Bede wrote two treatises, *On the Division of Time* and *On the Reckoning of Time*, which were concerned with calendar reckoning and in which he discussed such topics as chronology, astronomy, calendrical computations, Easter tables, and the tides. Although he borrowed heavily from his predecessors, especially Isidore, Bede was capable of adding intelligently to his meager inheritance. For example, he formulated the concept of the "establishment of the port" and recorded that the tides recur at approximately the same time at a particular place along the coast, although the times of occurrence vary from place to place.

## THE SEVEN LIBERAL ARTS

In a few instances, I have mentioned the seven liberal arts, and it will be useful to describe them more fully. They embraced both verbal and mathematical disciplines. The former, known as the "trivium," included grammar, rhetoric, and logic (or dialectic), whereas the latter, the "quadrivium," encompassed the four disciplines of arithmetic, geometry, astronomy, and music. All of these disciplines took form in classical Greece during the fifth and fourth centuries B.C., when they were first conceived as liberal arts suitable for teaching to free young men. The number of disciplines varied, however, and did not assume the canonical number of seven until the time of the Latin encyclopedists, who also coined the terms "trivium" and "quadrivium." The Latin encyclopedists shaped the seven liberal arts into the form they would have in the later Middle Ages. Martianus Capella's *Marriage of Philology and Mercury* was perhaps the quintessential Latin treatise that shaped the seven liberal arts. The setting of the book is the marriage of Mercury and Philology, where each of the seven arts is represented by a bridesmaid, who describes the art she represents. Others also wrote on the seven liberal arts, including Saint Augustine, Boethius, Cassiodorus, and Isidore of Seville. It was Cassiodorus who urged that the seven liberal arts be incorporated into a Christian education. By the end of the seventh century, the seven liberal arts were considered to be the basis of a proper education.

If there was such a thing as a core of scientific learning, it would be embedded in the quadrivium. Indeed, the four mathematical sciences that comprised it (arithmetic, geometry, astronomy, and music) were given their final condensed form by the Latin encyclopedists. Of the various accounts that discussed the quadrivium, the most popular and representative was Isidore of Seville's lengthy *Etymologies*. As the title suggests, Isidore was often concerned with etymological derivations of key terms, believing that knowledge of the origin of a term conveys an insight into the essence and structure of the thing it represents.

Isidore called attention to the importance of arithmetic for a proper understanding of the mysteries of Holy Scripture. For the arithmetic itself, he drew heavily upon Cassiodorus, who had, in turn, excerpted from the lengthy Boethian translation of Nicomachus's *Introduction to Arithmetic*. Isidore considered the division of numbers into even and odd and distinguished various subdivisions within each category. He enunciated a mélange of Pythagorean definitions, including those for excessive, defective, and perfect numbers (that is, where the sum of the factors of a number respectively exceeds, is less than, and equals the number itself), as well as for discrete, continuous, lineal, plane, circular, spherical, and cube numbers. If we add to these the definitions of five types of ratio distinguished by Nicomachus, then we have virtually the whole of Isidore's arithmetic. Faced with an unrelated collection of inept definitions, supplemented by a few trivial examples, the reader of Isidore's section on arithmetic could have used little of it. A comparison with the arithmetic books of Euclid's *Elements* (Books VII to IX) illustrates the depths to which arithmetic had fallen.

Isidore has even less to say about geometry. He begins with a strange fourfold division into plane figures, numerical magnitude, rational magnitude, and solid figures, and he concludes with definitions of "point," "line," "circle," "cube," "cone," "sphere," "quadrilateral," and a few others. Here we find "cube" defined as "a proper solid figure which is contained by length, breadth, and thickness," a definition applicable to any other solid (Euclid defined "cube" as a "solid figure contained by six equal squares"). A quadrilateral figure is "a square in a plane which consists of four straight lines," thus equating all four-sided figures with squares![10]

Isidore's longest section within the quadrivium is devoted to astronomy (the music section, like that on geometry, consists of a brief sequence of definitions). In a nontechnical description, Isidore considered the difference between astronomy and astrology, the general structure of the universe, the sun, the moon, the planets, fixed stars, and comets. We learn that the sun, which is made of fire, is larger than the earth and the moon; that the earth is larger than the moon; that, in addition to a daily motion, the sun has a motion of its own and that it sets in different places; that the moon receives its light from the sun and suffers eclipse when the earth's shadow is interposed between it and the sun; that the planets have a motion of their own; and that the stars, fixed and motionless in the heavens, are carried around by a celestial sphere, although the stars themselves are ranged at varying distances from the earth, an inference drawn from the observed unequal brightnesses of stars. Isidore believed that some of the more remote and smaller stars were actually larger than the bright stars that humans observe, their apparent smallness being merely a consequence of distance. It is unlikely that Isidore was aware that an unimaginably thick and transparent sphere would be required to accommodate fixed stars varying in size and distance and distributed under the conditions he described. Comprised, for the most part, of elementary and sketchy details, Isidore's astronomical discussion nevertheless represents his best effort among the subjects of the quadrivium.

The extent to which the quadrivial sciences were actually taught is problematic. It is not likely that they were taught in more than a cursory fashion. Few knew the four subject areas well enough to teach them, although Boethius had written treatises on arithmetic and music and perhaps also on astronomy and geometry. Nevertheless, the seven liberal arts served as an ideal core of education in the cathedral schools of the eleventh and twelfth centuries, and their study intensified in the twelfth century, even as new intellectual riches began to enter Europe from the Islamic world. With the emergence of universities in the late twelfth and thirteenth centuries, the liberal arts were greatly expanded as these growing institutions absorbed the new learning. Ironically, they were no longer taught as the seven liberal arts but rather as independent subjects in a broad-ranging curriculum. Indeed, the new emphasis at the universities was on natural philosophy and theology to which the liberal arts, insofar as they existed at all, were preparatory subjects.

The Latin encyclopedists supplied the early Middle Ages with most of what its scholars would know of science and natural philosophy. Their information was largely derived from the Greek and Latin handbook traditions. Too often, they failed to comprehend the material they read;

nonetheless, they copied it, or paraphrased it, in their own treatises. Despite their failings, the encyclopedists performed a vital service. Without their contributions, even the meager knowledge of the world that they provided would have been absent.

The encyclopedists provided late ancient and early medieval society with what has been characterized as "popular" science. Today, we also have popular science, ranging in quality from poor to excellent. A critical difference between our society and that of the Roman West is that the experimental and theoretical science on which our popular science is based was absent from Roman science during late antiquity and the early Middle Ages. Popular science in the Roman West was nearly coextensive with the whole of science. It was embodied in the quadrivial subjects described by the encyclopedists, who deserve our gratitude for their efforts to preserve the remnants of ancient science. There is no denying, however, that a scientific dark age had descended upon Western Europe.

# 2

## The new beginning: The age of translation in the twelfth and thirteenth centuries

THE division of the Roman Empire from the late third century onward into an essentially Greek-speaking eastern part and a Latin-speaking western part had a momentous consequence for intellectual life and therefore for the history of science and natural philosophy. With the passage of time, knowledge of the Greek language in the Western empire became relatively rare. Because Greek had been the language of science, this meant that Greek science was essentially unavailable to those whose sole language was Latin. To become accessible to the Latin-speaking West, a Greek scientific treatise had to be translated into Latin. Few treatises were. Apart from a small number of Hippocratic medical treatises and the few translations made by the likes of Chalcidius and Boethius, almost no significant works of Greek science were translated into Latin. During the ninth and tenth centuries, when the Arabs were translating a large portion of Greek science into Arabic and also adding to this legacy, and during the period when Greek science continued to be read and studied in the Greek-speaking Byzantine Empire, the West had before it only the rudimentary science of the Latin encyclopedists already described in chapter 1. By A.D. 500, knowledge of Greek had become rare and knowledge of the exact sciences even rarer. Except for occasional translations, which sometimes failed to circulate or perished completely, little was added to the now dominant encyclopedic tradition. Before Western Europeans seriously sought out new learning from neighboring civilizations and cultures, they first had to be aroused and stimulated to a new interest in science and nature.

For various reasons – including civil strife over the imperial succession that resulted in an empire split into western and eastern halves, economic deterioration because of waning trade and crushing taxes, and the massive migrations and invasions of Germanic and Celtic peoples into areas formerly dominated by Rome – most urban centers in Western Europe were in serious decline from approximately the fourth to the ninth centuries, a period that embraces the late Roman Empire and early Middle Ages. With the decline of the city, education and learning retreated to a considerable extent into the great and small monasteries that blossomed in the rural areas of Europe. Nevertheless, bishops in the various towns and cities still had to educate their clergy, and, for this purpose, some established schools under their own jurisdiction. In the late eighth century, Charlemagne (Charles the Great), who ruled from 768 to 814, mandated that all cathedrals and monasteries establish schools to educate the clergy, a difficult assignment as long as Scandinavian and other invaders continued to wreak havoc over much of the continent during the ninth and tenth centuries.

With the barbarian invasions finally over by the eleventh century – the Vikings were the last – a new Europe was emerging with new institutions, technologies, and ideas. Significant improvements in agriculture allowed for the feeding of a much greater population and, along with a dramatic increase in trade, led to a revival of urban life. During the eleventh and twelfth centuries, the cathedral schools in many European cities – Paris, Orleans, Toledo, Chartres, Cologne, and so on – became intellectual centers that attracted students and teaching masters. In these schools, future clergy learned Latin, the language of the Church. They also learned enough arithmetic to perform calculations with Roman numerals and to cope with calendrical problems, both secular and religious. To some degree, they were taught the rudiments of the seven liberal arts, and they even studied classical Latin literature through which they were exposed to a broader cultural history. From the tenth to the twelfth centuries some great teaching masters emerged, who attracted students from all over Europe and trained other teaching masters. Among the most famous were Gerbert of Aurillac (ca. 946–1003), who became Pope Sylvester II (999–1003), Adalberon of Laon, John of Auxerre, Thierry of Chartres, Fulbert of Chartres, Peter Abelard, William of Conches, Clarenbald of Arras, and John of Salisbury.

As one of the earliest and best known cathedral school masters, Gerbert reveals considerable breadth of interest when he used Church contacts in northern Spain to acquire a few Arabic treatises in Latin translation. From these he learned about the abacus and the astrolabe, writing a treatise on the former and perhaps also on the latter. The substance of his work falls squarely in the Latin tradition. Gerbert was not an original thinker, however, and his subsequent influence was based largely on his talents as a teacher of science. From 972 to 989, he taught the seven liberal arts at the cathedral school of Reims, where he emphasized a rather elementary mathematics and astronomy. Gerbert even used visual aids, which helped him gain a reputation as a great teacher in an age of intellectual deprivation. He not only explained how to construct a sphere to represent the heavens, he actually made one. Gerbert's sphere simulated the motions of the constellations, using wires fixed on its surface to outline the stellar configurations. Deeply impressed with Gerbert's ingenuity and dedication, his pupils went forth enthusiastically to continue and extend his teachings, emphasizing science as an integral part of the liberal arts. Many of the cathedral schools that rose to prominence in the eleventh and twelfth centuries, replacing the monastic schools as centers of learning, were either founded or revived by Gerbert's pupils. Gerbert's most eminent students were Adalberon of Laon, John of Auxerre, and especially Fulbert of Chartres.

Despite the lack of coherent and challenging scientific texts, the cathedral school environment encouraged intellectual interest in secular and scientific subjects. There is ample evidence of this in an extraordinary exchange of eight letters on mathematics, sometime around 1025, between two cathedral school products, Ragimbold of Cologne and Radolf of Liège. At the initial request of Radolf, a series of mathematical questions was posed and the answers were circulated not only to the two correspondents but also to others who seem to have acted as judges in what may aptly be described as a scientific tournament. Ragimbold and Radolf had only a meager and fragmentary knowledge of geometry. Because of their ignorance of Greek and Arabic mathematics, they were dependent on tidbits of geometry drawn from Roman surveying manuals and on the genuine and dubious writings of Boethius. Neither had any concept of geometric demonstration. Among other things, we find an utterly confused discussion on the meaning of exterior and interior angles in a triangle. In a question drawn from Boethius's commentary on the *Categories* of Aristotle, Radolf asked Ragimbold to calculate the side of a square that is double a given square. Both knew that the side of the larger square is the diagonal of the smaller given square (see Figure 1). But our contestants were unaware that the sides of the two squares could not be related by such whole number ratios as $17/12$ (Ragimbold's ratio) or $7/5$ (Radolf's), because the two sides are incommensurable and therefore relatable only by an irrational ratio, $2/1$ in this case. The lack of mathematical comprehension in this contest is of less significance than the fact that such a tournament took place at all. Its occurrence signifies a growing interest in scientific questions, and it would probably not have happened one hundred years earlier.

## EDUCATION AND LEARNING IN THE TWELFTH CENTURY

The positive spirit exemplified by Ragimbold and Radolf toward mathematics in the eleventh century found its counterpart in natural philosophy one century later. Plato's *Timaeus* (and a few commentaries on it) and the literary legacy of the Latin encyclopedists provided a more substantial literature to contemplate in natural philosophy than Ragimbold and Radolf had had available to them in mathematics. The level of scientific fare was not, however, the decisive factor in producing intellectual changes in the twelfth century. Somehow, a remarkable change of atti-

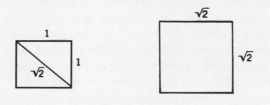

*Figure 1*

tude occurred toward traditional authorities and toward nature itself. Whatever the societal reasons for these changes, which will be described in the next chapter as part of the background for the emergence of universities, the idea that God was the direct and immediate cause of everything yielded to an interpretation of the world that assumed that natural objects were capable of acting upon each other directly. God had conferred on nature the power and ability to cause things. He had made of it a self-operating entity. Nature, or the cosmos, was thus objectified and conceived as a harmonious, lawful, well-ordered, self-sufficient whole, which could be investigated by the human intellect. The world was transformed conceptually from an unpredictable, fortuitous entity to a smoothly operating machine, or *machina*, as it was frequently called in the twelfth century. What developed was the concept of a "common course of nature" in accordance with which nature operated routinely and regularly. Nature's ordinary activities could be suspended only by divine intervention, which was sometimes interpreted as part of the divine plan of which we humans are ignorant. Theologians who were tradition minded found this newly awakened interest in the operations of nature threatening. A typical representative of the old order, Absalom of Saint-Victor, denounced what he viewed as ceaseless inquiries into "the composition of the globe, the nature of the elements, the location of the stars, the nature of animals, the violence of the wind, the life-processes of plants and of roots."[1] William of Conches spoke for most of his like-minded colleagues against the traditionalists when he declared that

> Ignorant themselves of the forces of nature and wanting to have company in their ignorance, they don't want people to look into anything; they want us to believe like peasants and not to ask the reason behind things. . . . But we say that the reason behind everything should be sought out. . . . If they learn that anyone is so inquiring, they shout out that he is a heretic, placing more reliance on their monkish garb than on their wisdom.[2]

William insisted that God's power is enhanced by assigning secondary causes, which not only enable nature to operate but through nature also produce the human body.[3] Those imbued with the new spirit of inquiry thought it incumbent upon the faithful to discover the laws of nature. Nature, or the cosmos, was an entity to be studied in order to better understand God's creation. In this noble task, however, the guide was philosophy, not the Bible. God was to be invoked as an explanatory cause only when natural causes were to no avail. In the annals of Christianity, the power of reason was exalted as never before. The quest for secondary causes in nature placed an emphasis on the natural order and its lawful operations. Secular learning acquired prestige and was viewed by some as a challenge to theology and biblical explanation. A new era had dawned in which natural inquiry played a role, as exemplified by the writings of William of Conches, Honorius of Autun, Bernard Silvester, Adelard of Bath, Thierry of Chartres, and Clarenbald of Arras.

The new attitude toward nature emerged from within the old Latin learning. It was based largely on the works of the Latin encyclopedists, Plato's *Timaeus* (the part translated into Latin) and the various commentaries on it, the *Division of Nature* (*De divisione naturae*) by John Scotus Eriugena (ca. 810–ca. 877), a few traditional Latin literary works, and other Latin treatises written in the tenth and eleventh centuries. If given sufficient time and left relatively unimpeded, twelfth-century scholars might have produced a significant long-term tradition of science and natural philosophy. But the momentous influx of science and natural philosophy from Islam had already begun and would soon overwhelm the incipient rational science that had been evolving within the context of the old learning.

## LATIN TRANSLATIONS FROM ARABIC AND GREEK

The new concern for nature and its operations generated an intense interest in the works of Greek antiquity, many of which were available only in Arabic translation. Thus to the Greek legacy must be added the contributions of numerous Islamic authors, a group that includes not only Muslims but also Christians and Jews, who were bound together by their common use of the Arabic language. The works of these scientists, natural philosophers, and physicians were largely unknown to twelfth-century European scholars. The totality of this body of literature in both the Greek and Arabic languages is what is commonly referred to as the Greco-Arabic (or Greco-Islamic) heritage. The desire for Greco-Arabic learning grew out of an almost worshipful respect for ancient learning and wisdom as twelfth-century scholars recognized an incalculable debt to their predecessors. If they envisioned an extension of the horizons of knowledge, it was only because, as Bernard of Chartres expressed it, they were privileged to stand upon the shoulders of the learned giants of antiquity, a sentiment repeated often through the centuries and even found in a letter by Isaac Newton. But the works of those giants had been either unavailable or known only in fragments. Reports of treatises that existed in either Greek or Arabic, but that were known in the West only by title, or not at all, aroused the curiosity and desires of Western scholars while simultaneously reinforcing a sense of intellectual deprivation. To remedy this serious deficiency, scholars of the Western world acted to acquire the scientific heritage of the past. They began to translate treatises from Arabic and Greek into Latin because, as their prefaces often inform us, they wanted to present the treasures of the East to the West and thus relieve the "poverty of the Latins" (*Latinorum penuria*) in so many fields. Their translations constitute one of the true turning points in the history of Western science and natural philosophy.

Already in the middle of the tenth century, translations from Arabic to Latin were made in northern Spain at the Monastery of Santa Maria de Ripoll at the foot of the Pyrenees. These translated works were largely concerned with geometry and astronomical instruments and were perhaps known directly by Gerbert. In the eleventh century, information about the Arabic astrolabe was known to Hermann of Reichenau (1013–1054) and translations of medical treatises of Greek and Arabic authors were made from Arabic to Latin by Constantine the African (fl. 1065–1085), a shadowy figure associated with the medical center at Salerno in southern Italy. But the translating activity that revolutionized Western scientific thought and determined its course for centuries to come occurred in the twelfth century and, to a lesser extent, in the thirteenth. Between 1125 and 1200, a veritable flood of translations into Latin made a significant part of Greek and Arabic science available, with more to come in the thirteenth century. Not since the ninth and early tenth centuries, when much of Greek science was translated into Arabic, had anything comparable occurred in the history of science.

The great age of translation was preceded by the rollback of the Muslims in Spain and their defeat in Sicily during the eleventh century. With the fall of Toledo in 1085 and the capture of Sicily in 1091, a now reinvigorated Western Europe came into possession of significant centers of Arabic learning. Books in Arabic were readily at hand, and intellectually starved Europeans were eager to make their contents available in Latin, the universal language of learning in Western Europe. Scholars came from all parts of Europe to join with native-born Spaniards, whether Christian, Jew, or Muslim, to engage in the grand enterprise of converting technical science and natural philosophy from the Arabic language into Latin, a language that had hitherto been largely innocent of such matters. The international character of this extraordinary activity is revealed by the very names of the most significant translators, among them Plato of Tivoli, Gerard of Cremona, Adelard of Bath, Robert of Chester, Hermann of Carinthia, Dominicus Gundissalinus, Peter Alfonso, Savasorda, and John of Seville. In the early thirteenth century came Alfred Sareshel (or Alfred the Englishman), Michael Scot, and Hermann the German.

Among Spanish centers where translations were made, Toledo was foremost. There, and elsewhere, translations were made in a variety of ways. If the translator had mastered Arabic adequately, he could translate directly; if not, he might team with an Arab or a Jew. Occasionally, if he knew Spanish, he might engage someone to translate from Arabic to Spanish and himself translate from the latter to Latin. A Latin translation of an original Greek treatise might occasionally be converted through a sequence of languages, say from Greek to Syriac to Arabic to Spanish to Latin or, perhaps, from Arabic to Hebrew to Latin. Some distortions in the Latin end product from successive translations could hardly have been avoided, although the overall results were quite satisfactory, especially with regard to the works of Aristotle.

The translations of the twelfth and thirteenth centuries were overwhelmingly of scientific and philosophical works – the humanities and belles lettres were scarcely represented. The manner in which works were selected for translation was frequently haphazard. Availability and brevity were often the decisive factors. Treatises of genuine significance were sometimes ignored, whereas minor and occasionally trivial works were translated and subsequently studied with great intensity. Because the translators worked in widely separated locations and were rarely in contact, duplication of effort was all too common. Nevertheless, despite great obstacles, the sum total of achievement is impressive. Indeed, translations by Gerard of Cremona (d. 1187) alone drastically altered the course of Western science. In a tribute to this greatest of all Western translators and as a guarantee that posterity would recognize its indebtedness to him, and not credit his accomplishments to others, Gerard's devoted students appended a biographical sketch and a list of his translations to Gerard's translation of Galen's *Tegni* (*Medical Art*). Here we learn that after absorbing all that was available to the Latins, Gerard went to Toledo to find Ptolemy's *Almagest*, which could not be found among the Latins. "There [in Toledo], seeing the abundance of books in Arabic on every subject, and regretting the poverty of the Latins in these things, he learned the Arabic language, in order to be able to translate." "To the end of his life," we are told, Gerard "continued to transmit to the Latin world (as if to his own beloved heir) whatever books he thought finest, in many subjects, as accurately and as plainly as he could."[4] Not only did Gerard translate the *Almagest*, but he also translated at least seventy other treatises. Among these were the basic physical works of Aristotle (*Physics, On the Heavens and World, On Generation and Corruption*, and *Meteorology*, Books I to III), as well as Aristotle's *Posterior Analytics*, the major treatise for discussion of scientific method. Gerard also translated numerous mathematical works, including Euclid's

*Elements,* the *Algebra* of al-Khwarizmi, and *The Geometry of the Three Brothers,* which contained significant, and subsequently influential, Archimedean mathematical techniques. In addition to other astronomical, astrological, alchemical, and statical works, Gerard translated a large number of medical treatises, including many by Galen, as well as the *Canon of Medicine* of Avicenna and the *Liber Continens* (that is, *The Book of Divisions Containing 154 Chapters*) of Rhazes (al-Razi). These works formed the core of medieval medical studies.

Significant translations were also made directly from Greek to Latin. These were done almost exclusively in Italy and Sicily, where contacts with the Greek-speaking Byzantine Empire had never been broken. During the twelfth century, the Norman rulers of southern Italy and Sicily utilized these contacts to gather Greek theological, scientific, and philosophical texts. In Sicily, Plato's *Meno* and *Phaedo* were translated (by Henricus Aristippus), as was Ptolemy's *Almagest,* Euclid's *Optics, Catoptrics,* and *Data,* and a few of Aristotle's works. Translations were also made from Arabic. Eugene the Emir, who was trilingual (Arabic, Greek, and Latin), translated Ptolemy's *Optics* from Arabic to Latin. In northern Italy, where the names of James of Venice, Burgundio of Pisa, and Moses of Bergamo are preserved, additional translations were made from Greek to Latin. But just as Gerard of Cremona towered above all other translators from Arabic to Latin, so William of Moerbeke (ca. 1215–ca. 1286), a Flemish Dominican, was supreme over all translators from Greek to Latin. Encouraged by his friend, Saint Thomas Aquinas, who had complained of the inadequacy of the translations of Aristotle's works from Arabic, Moerbeke completed new translations from Greek manuscripts of almost all of Aristotle's works except the *Prior* and *Posterior Analytics.* To these, he added translations of commentaries on Aristotle's works by some of the most important Greek commentators of late antiquity, such as Alexander of Aphrodisias, John Philoponus, Simplicius, and Themistius. In 1269, he translated all but a few of the numerous works of Archimedes, along with important Greek commentaries. Renaissance translators utilized these translations without acknowledgment and inadvertently paid Moerbeke high tribute by publishing his translations in the first printed version of the works of Archimedes in 1503 at Venice. Moerbeke made at least forty-nine translations ranging over theology, science, and philosophy.

Methods of translation varied considerably. Sometimes a translator conveyed little more than the sense of a treatise. Usually, however, translators sought to capture the substance of a work, while also preserving the sense of the words. The most frequent method for achieving this end was by a word-for-word translation (*verbum de verbo*), a method that worked far better for translations from Greek than from Arabic, because the former language was structurally similar to Latin, whereas the latter was not. Aware of this, medieval scholars preferred translations from the Greek whenever possible. Medieval translators paid little attention to literary style. They preferred a more praiseworthy objective: fidelity to the original text.

### THE TRANSLATION OF THE WORKS OF ARISTOTLE

Because the interpretation and comprehension of Aristotle's natural philosophy in medieval Western Europe looms large in this study, it will be useful to say something about the manner in which his works were translated into the Latin language. Of the translations of Aristotle's works, those made directly from Greek were far more numerous than those made from Arabic. If we confine ourselves to five of Aristotle's "natural books" (that is, the books primarily concerned with natural philosophy), say the *Physics, On the Heavens, On Generation and Corruption, Meteorology,* and *On the Soul,* and compare the number of extant manuscripts that were translated from Greek originals to those translated from Arabic originals, the figures are striking. For the *Physics,* we obtain 371 extant manuscripts translated from Greek, and 134 from Arabic; for *On the Heavens,* 190 from Greek, 173 from Arabic, the closest comparison; for *On Generation and Corruption,* 308 from Greek, 48 from Arabic; for *Meteorology,* 175 to 113; and for *On the Soul,* 423 to 118. For only one of these five treatises is the difference small. For the other treatises, the disparity between the two sources is obvious. Where Greek manuscripts for a text were available, they were always preferred for the reasons already given. Not only was the structure of the Arabic language radically different from that of Latin, but some Arabic versions had been derived from earlier Syriac translations and were thus twice removed from the original Greek text. Word-for-word translations of such Arabic texts could produce tortured readings. By contrast, the structural closeness of Latin to Greek, permitted literal, but intelligible, word-for-word translations.

The primary translators of Aristotle's works from the Greek were Boethius in the early sixth century; James of Venice, Henricus Aristippus, and Ioannes, a little-known figure, in the twelfth century; and Robert Grosseteste and William of Moerbeke in the thirteenth century, the latter undoubtedly the greatest of all translators from Greek into Latin. Aristotle's major translators from Arabic to Latin were Gerard of Cremona in the twelfth century, who translated most of Aristotle's natural books, and Michael Scot in the thirteenth century, who translated Aristotle's biological treatise, *On Animals (De animalibus).*

According to scholars, approximately two thousand Latin manuscripts of the works of Aristotle have been identified. If this number of manuscripts survived the rigors of the centuries, it is plausible to suppose that thousands more have perished. The extant Latin manuscripts are a good measure of the pervasive hold that the works of Aristotle had on the intellectual life of the Middle Ages and Renaissance. With the possible exception of Galen (ca. 129–ca. 200), the great Greek physician of late antiquity, no other Greek or Islamic scientist has left a comparable manuscript legacy.

Before leaving the translations, we might ponder whether major and sustained translating activity could have occurred earlier, say in the tenth or the eleventh or in both centuries. Suitable conditions for large-scale translations in those earlier centuries appear to have been absent. The Arabs did not themselves complete their translations of Greek science and natural philosophy into Arabic until the tenth century. Moreover, the translations were made in the eastern part of Islam, especially in Baghdad. The majority of these texts may not have reached the Arabs in Spain, Sicily, and southern Italy until perhaps the eleventh century. Just at that time, Christians were active in the reconquest of Spain and Sicily, producing unfavorable conditions for major translating activity until the capture of Toledo in 1085 and the Norman conquest of southern Italy and Sicily by the end of the eleventh century. In that century, in Salerno, Italy, Constantine the African translated some medical treatises from Arabic into Latin. Translations on a grand scale, however, awaited the twelfth century in Spain, largely between 1140 and 1160. Thus it was not likely that the translations could have occurred earlier than the twelfth century, because the texts were not readily available and the Christian–Muslim conflicts in Spain and Sicily were too intense until the late eleventh century. By that time, however, the Christians had advanced far enough into Spain to allow for more settled conditions and had by then, or shortly after, come into direct contact with the large body of Arabic scientific literature.

## THE DISSEMINATION AND ASSIMILATION OF ARISTOTLE'S NATURAL PHILOSOPHY

The introduction of the works of Aristotle into the Latin language and the subsequent dissemination and assimilation of those works transformed the intellectual life of Western Europe. But Aristotle's influence did not depend solely on his own works.To assess the enormous impact of Aristotle, we must consider the commentaries on his works that were composed by Greeks in late antiquity and by Arabs during the ninth to twelfth centuries. Although Aristotle's genuine works shaped the medieval perception of the world, many works that were falsely ascribed to him also shaped medieval judgments about his views. To these we must add Latin translations from the Arabic of non-Aristotelian treatises that contained ideas derived from Aristotle's natural philosophy, especially in medicine and astrology. This massive complex of Aristotelian ideas and interpretations was inherited by the natural philosophers of the Latin Middle Ages. Utilizing these sources, medieval scholars proceeded to add their own commentaries on Aristotle's works, as well as to compose specialized treatises in which Aristotle's ideas were paramount. The totality of this body of literature – the inheritance and the additions thereto – is what we today call "Aristotelianism." This term, which was never used in the Middle Ages, admirably characterizes the major component of intellectual life from the twelfth century to the fifteenth (the Middle Ages proper) and even beyond to the end of the seventeenth century.

## The contributions of Greek commentators

Through commentaries on the works of Aristotle, the Greek world of late antiquity contributed significantly to natural philosophy. Working between A.D. 200 and 600, Greek commentators left behind numerous treatises, the extant part of which comprises approximately fifteen thousand pages of Greek text in the edition known as the Ancient Greek Commentaries on Aristotle (*Commentaria in Aristotelem Graeca*). Of the authors who commented on Aristotle, some were Aristotelians and others Neoplatonists, who were quite critical of Aristotle. Of this group, those whose influence on Islamic and Latin science and philosophy was most extensive were Alexander of Aphrodisias (fl. 198–209), Themistius (fl. late 340s–384/385), Simplicius (ca. 500–d. after 533), and John Philoponus (ca. 490–570s), a Neoplatonist who was also a Christian. The influence of Alexander and Themistius on natural philosophy in the Latin Middle Ages came largely through the Aristotelian commentaries of Averroes, the famous Muslim commentator, who frequently cited passages from their works. Simplicius's commentary on Aristotle's *On the Heavens* (*De caelo*), which William of Moerbeke translated into Latin in the thirteenth century, conveyed important ideas on cosmology and physics. Although most of John Philoponus's works remained unknown in the Latin West until the sixteenth century, some of his ideas were known through William of Moerbeke's partial translation of his commentary on Aristotle's *On the Soul*, through Simplicius's attacks upon him in the former's commentary on Aristotle's *On the Heavens*, and through occasional citations of his ideas in Averroes's Aristotelian commentaries. Philoponus is important in the history of science because he was critical of Aristotle's ideas in physics and cosmology. Impetus theory, or the doctrine of impressed force, which played a significant role in Arabic and medieval Latin physics, was derived ultimately from Philoponus's commentary on Aristotle's *Physics*. He also insisted, against Aristotle, that finite motion was possible in a vacuum and that two unequal weights dropped from a given height would strike the ground almost the same time. In his commentary on Genesis (the *De opificio mundi*), he argued against Aristotle's concept of the eternity of the world and also insisted that celestial and terrestrial matter are identical rather than radically different, as Aristotle had claimed. In recent years, the significance of the late Greek commentators has been much more appreciated, and their contributions to the history of medieval and early modern science may ultimately prove greater than once thought.

## The contributions of Islamic commentators

When Aristotle's works were translated from Greek (or even Syriac) to Arabic during the ninth and tenth centuries, it was not long before Islamic scholars studied those works and wrote commentaries on them. Islamic commentaries and discussions of Aristotle's works and ideas that influenced the West were written prior to 1200. Because a number of Neoplatonically inspired Greek commentaries on Aristotle had been translated into Arabic, Neoplatonic ideas were often incorporated into Islamic commentaries on Aristotle. Among Muslim scholars who wrote on Aristotle in Arabic and who had works translated into Latin, the most important were al-Kindi (ca. 801–ca. 866), al-Farabi (ca. 870–950), Avicenna (Ibn Sina) (980–1037), al-Ghazali (1058–1111), and Averroes (Ibn Rushd) (1126–1198). Of this group, Avicenna, al-Ghazali, and Averroes had the greatest impact on Aristotelian natural philosophy in the West. The most influential Jewish scholar in Islam to contribute to European scholarship was Moses Maimonides (1135–1204), who wrote in Arabic.

In his lengthy *Kitab al Shifa (The Book of the Cure [of Ignorance])*, a philosophical encyclopedia translated in the twelfth century by Dominicus Gundissalinus and Avendauth (Abraham ibn Daud), Avicenna commented upon many aspects of Aristotle's natural philosophy. The second part of that work was devoted to physics, which, in the incomplete twelfth-century Latin translation, was called *Sufficientia* and consisted of eight parts. In the sections that medieval natural philosophers had available, they could read Avicenna's ideas about the heavens, generation and corruption, the elements, meteors, animals, minerals, and the soul. His great medical work, *Canon of Medicine*, may have been more important in the medical schools of the medieval universities than were the works of Galen.

Although al-Ghazali had a significant impact on the West, it was not because of his own views and interpretations. Al-Ghazali had written a summary of the philosophical opinions of al-Farabi and Avicenna and subsequently wrote a severe criticism of their views. Only the former was translated into Latin. In this way, the opinions of al-Farabi and Avicenna were attributed to al-Ghazali. His untranslated criticism of philosophy, *The Incoherence of the Philosophers*, became known in the West from the critique of it in Averroes's *The Incoherence of the Incoherence*, which was translated into Latin.

Among all Islamic authors, Averroes was the one who most influenced the Aristotelian outlook of the Latin West. An eminent scholar has observed that "If there is a process of naturalization in literature corresponding to that in citizenship, the writings of Averroes belong not so much to the language in which they were written as to the language into which they were translated and through which they exerted their influence upon the course of the world's philosophy."[5] It is one of history's ironies that Averroes's Arabic works were virtually ignored by Arabic-speaking peoples in Islamic countries, but that many of those same works exerted a great influence in Christendom by way of Latin translations.

Thus far some thirty-eight Arabic commentaries by Averroes on the works of Aristotle have been identified. This large number results from the fact that Averroes wrote at least two, and often three, different kinds of commentaries on any given Aristotelian treatise. On Aristotle's *Physics*, for example, he wrote an epitome, or brief summary; a middle commentary, or paraphrase of the text; and a long commentary, which was a detailed, sequential discussion of the successive sections of the entire text. He applied this same threefold treatment to Aristotle's *On the Heavens* and *Metaphysics*. For some of Aristotle's treatises – *On Generation and Corruption* and *Meteorology*, for example – he wrote only middle and long commentaries. Of the thirty-eight Arabic commentaries, fifteen were translated from Arabic into Latin during the first part of the thirteenth century (by Michael Scot and others), and nineteen more were translated from Hebrew into Latin during the sixteenth century (Averroes's commentaries were even more influential in the Jewish Aristotelian tradition than in the Latin tradition). In his commentaries, Averroes sought to purge Aristotelian thought of the Neoplatonic interpretations that had, in his view, distorted Aristotle's true meaning. He was convinced that Aristotle had grasped as much truth about the world as was possible for a human being using demonstrative proof.

### Pseudo-Aristotelian works

Beginning some two generations after Aristotle's death, the attribution of spurious works to the philosopher began with two Greek titles, *On Colors (De coloribus)* and the *Mechanics (Mechanica)*. In time, other spurious Greek works were added. This, however, was only the beginning. The process of false attribution was repeated in every language into which Aristotle's works were translated, which included Syriac, Arabic,

Latin, Hebrew, Armenian, and some European vernacular languages. Most of the spurious works were in the area of pseudoscience, especially in alchemy, astrology, chiromancy, and physiognomy. Astronomy was also represented. Many of these spurious works were translated into Latin from Arabic. In the Latin world, most of them circulated independently of Aristotle's genuine works. They seem to have appealed to a different social group than that in the university environment, where, with a few exceptions, they had little impact and were rarely cited in works on natural philosophy. Among the exceptions are the *Book of Causes* (*Liber de causis*, translated by Gerard of Cremona), which was based on Proclus's *Elements of Theology* and was especially influential among theologians, eliciting commentaries from Albertus Magnus and Thomas Aquinas; *On the Causes of the Properties of the Elements* (*De causis proprietatibus elementorum*), which appears in numerous manuscript codices of Aristotle's natural books and exerted a major influence in the thirteenth and fourteenth centuries; and, finally, although less important for natural philosophy than the first two treatises, the *Secret of Secrets* (*Secretum secretorum*), which presents a large number of maxims that ostensibly encapsulate the wisdom that Aristotle was said to have transmitted to ancient rulers. Of all the spurious works attributed to Aristotle, the *Secret of Secrets* was the most popular, as evidenced by at least six hundred extant manuscripts of which twenty or so circulated with one or more of Aristotle's genuine works.

## RECEPTION OF THE TRANSLATIONS

The texts of Aristotle were difficult and the translations not always clear, occasionally prompting charges of obscurity. The commentaries of Avicenna and Averroes were thus enthusiastically welcomed as guides for the interpretation of Aristotle's demanding texts.

Aristotle's influence in Western thought began even before the massive translations, largely because of two Latin translations of Abu Ma'shar's Arabic treatise on astrology, one in 1133, the other in 1140. Abu Ma'shar's *Introduction to Astronomy* was an astrological work that incorporated numerous ideas and concepts from Aristotle's natural books. Many twelfth-century scholars first met the doctrines of Aristotle through Abu Ma'shar's treatise. But this trickle of isolated Aristotelian ideas was soon overwhelmed by the translations of his works. Despite the new translations of Aristotle's works in the twelfth century, few manuscripts survive from that period, indicating that Aristotle's treatises had little direct influence in that century. The situation is dramatically altered by the middle of the thirteenth century, however, by which time manuscripts of Aristotle's works turn up in large numbers. By then Aristotle's influence had become significant and would only increase with time. An important measure of his impact is the production of Latin commentaries on his works, a subject that will be treated in a later chapter.

Virtually all of the ancient Greek treatises translated into Latin from Greek or Arabic, or both, were previously unknown in Christian Western Europe. How was this large body of pagan science and natural philosophy received? How did Christians respond to a body of literature with which they were completely unfamiliar and which had potential problems for the faith? Although these treatises were new to Western Europe, experience with pagan literature was not. Christians had adjusted to it long before. They had been exposed to pagan thought almost from the moment that the Christian religion was disseminated beyond the Holy Land. Not only was pagan thought familiar in the Greek-speaking eastern part of the Roman Empire, but Latin authors in the West, such as Saint Augustine, Saint Ambrose, and the Latin encyclopedists, were also familiar with pagan ideas. Because of Christianity's previous experience with pagan literature, the Latin translations of Greco-Arabic science in the twelfth and thirteenth centuries may be viewed as a second, and much more extensive, influx of pagan thought to the Christians of Western Europe. Although the science and natural philosophy of the second wave of pagan thought caused some friction between faith and reason, Christian natural philosophers, many of whom were theologians, were delighted to receive them. With Aristotle's logic and natural philosophy as its centerpiece, the new learning furnished the curriculum of the newly emerging universities, which formed one of most enduring institutional legacies from the Middle Ages, and which we must now describe.

# 3

# *The medieval university*

A description of the structure and operation of the medieval universities is essential because of the importance of these institutions for the development of Western science. The universities had emerged as a result of the transformation of society and intellectual life that had occurred in Western Europe by the twelfth century.

The highly feudalized Europe of the seventh and eighth centuries was drastically altered by the eleventh century. During the late eleventh and twelfth centuries, political conditions improved dramatically, due in no small measure to French-speaking feudal lords who brought reasonably stable governments to Normandy, England, Italy, Sicily, Spain, and Portugal. The vigor of a revitalized Europe is also evidenced by the reconquest of Spain, which was well underway by the end of the eleventh century.

With the establishment of greater security, Europe's economy revived, and the standard of living rose for all segments of society. This was occasioned by significant agricultural improvements, most notably the advent of the heavy plough, to which the horse was now harnessed instead of the ox. This substitution was made possible by the introduction of the nailed horseshoe and the collar harness, which together made horses far more effective agricultural engines than oxen. No less significant was the replacement of the two-field system of crop rotation with the three-field system, which also allowed for a major increase in food production. An augmented food supply was instrumental in bringing about a considerable growth in the population, which, in turn, made possible an expansion of cities and towns. Indeed, the increased population eventually made it necessary to build hundreds of new towns. Europeans began to colonize previously unpopulated or underpopulated lands, or they drove eastward against the Slavs, as the Germans did in their movement beyond the Elbe River. In the Low Countries, they even began to reclaim land from the sea. Europeans were on the move and significant migrations occurred. Many of the new towns were populated by free men, often former serfs who had fled to the towns in hopes of a better life.

By the end of the twelfth century, the level of commerce and manufacturing in Europe was probably greater than it had been at the height of the Roman Empire. Between the ninth and thirteenth centuries, Europe was transformed. A money economy had come into being.

Changes in government were also in the offing. The struggle between the towns and cities, on the one hand, and secular and ecclesiastical rulers, on the other hand, was underway. Increasingly, urban populations sought as much self-government as they could get and struggled to free themselves from taxes imposed by hereditary nobles. The concept of a commune with its attendant rights of citizenship was developed. European cities opportunistically sided with popes, kings, emperors, or independent princes to increase their power and protect their rights.

Thus cities became a powerful force in the economic, political, religious, and cultural life of the European continent. Because the universities of Europe were urban creations, one might infer that they were somehow the product of the forces just described, but this would be incorrect. Cities were only a necessary, not a sufficient, condition for the emergence of universities. Urbanization may have provided an essential matrix within which universities could develop and flourish, but it was hardly a guarantee that the process would actually occur. From the earliest societies of ancient Egypt and Mesopotamia, numerous diverse urban civilizations had come and gone, but none had produced anything comparable to the universities of Europe. Indeed, universities are hardly essential for a civilization to reach a lofty state of intellectual achievement. To keep records, preserve literary traditions, and add to the storehouse of societal knowledge and wisdom, a civilization needs only to insure that some of its members can read and write, that enough of them perform the requisite tasks, and that the written record is somehow preserved and transmitted from generation to generation. Societies that have met these criteria have reached great intellectual heights, as the medieval civilizations of Islam and China amply demonstrate.

Although the Latin West derived its science and natural philosophy from the Greeks and Arabs, the university was an independent invention that grew from conditions peculiar to the West in the twelfth century. The growing commercial life in urban centers had made it advisable, if not necessary, for those who practiced the same trade or craft to seek protection by organizing themselves into guilds or corporations. Medieval lawyers frequently called these organizations by the name *universitas*, that is, "totality" or "whole," signifying that the guild in question represented all the rightful practitioners of that trade or craft.

Teaching masters and students formed a vital part of twelfth-century society. They established important schools in various cathedrals of Western Europe, especially at Paris, Chartres, and Orleans. Students and masters customarily moved from one school to another, the students searching for the right master, the masters seeking to attract sufficient students to provide them with appropriate remuneration. Most masters and students were foreigners in the cities in which they taught and studied and, consequently, had no rights and privileges. Operating individually, they were no match for the municipal, state, and church authorities with whom they had to negotiate teaching conditions.

The masters and students in Paris and elsewhere saw the advantages of association and used the *universitas* of a trade or craft as a model on which to base their own organization. By the end of the twelfth century, there were already de facto organizations of masters, students, or both known as "universities" (for example, *universitas magistrorum*, or "university of masters"; *universitas scholarium*, or "university of students"; and *universitas magistrorum et scholarium*, or "university of masters and students"). Eventually, the term *universitas* by itself was sufficient to identify an educational institution. Although many guilds and corporations had preempted the term *universitas* before educational institutions of higher learning adopted it, the latter came to retain the term permanently, perhaps because they outlasted all the others.

Because of its subsequent significance, the term "university" (*universitas*) requires further explanation. At the outset, it was applied to a single group that formed a legally recognized self-governing association. Thus a faculty of arts was a "university," as was any faculty of medicine or faculty of theology. The masters and students of the arts faculty formed their own legal corporation, or university, as did the teachers and students of the medical faculty, and so on. Many student associations were also recognized as universities, especially in Italy.

The term that was initially employed, and was in common use by the middle of the thirteenth century, to encompass all of these individual, disparate universities, or university associations, was *studium generale*. Every master and student was a member not only of his own individual university, or corporation, but also of the *studium generale*. Where a single faculty, or corporation, or even two, operated a school, the designation *studium generale* would not normally have been conferred upon it. The term was usually assigned to schools that either were sufficiently prestigious, such as the customary universities of Paris, Oxford, and Bologna, or were large enough to include at least three of the four traditional faculties (arts, theology, law, and medicine), or were both. A major advantage for a school designated a *studium generale* was a precious right automatically conferred on its graduates: the "license [or right] to teach anywhere," known in Latin as the *ius ubique docendi*. In practice, however, it was more the prestige of a *studium* that validated the right of its graduates to teach anywhere.

It is obvious that the term *studium generale* is equivalent to our modern term "university." Perhaps toward the end of the Middle Ages, "university'' replaced *studium generale* and became the term as we know it today and that we shall use henceforth.

As corporate entities, the various medieval guild associations received important monopolistic privileges. The universities were no exception, receiving special treatment from church and secular authorities, who sought to encourage their growth. Each faculty was given jurisdiction over its own internal affairs and therefore had the right to judge the worthiness of the teachers and students who joined it as corporate members. The university, comprised of its faculties and students, had the legal right to negotiate on a wide range of problems with the external authorities that controlled the various governmental and religious jurisdictions in which it was located. There were also privileges relevant to personal status. Members of the *universitas* were accorded certain crucial rights, the most important of which was clerical status. Although most masters and students were neither ordained nor in orders, clerical status conferred upon them the rights of clergy. To attack a student or master who was traveling was the same as attacking a priest and was subject to severe penalties. Clerical status also allowed students who were arrested by civil authorities to demand trial in ecclesiastical courts, which were usually more lenient than civil courts. It also permitted students and masters to receive benefices from the church and to enjoy the fruits of those benefices while pursuing their regular university activities. In addition to these individual privileges, an important corporate right allowed the universities to suspend lectures and even to depart from their respective cities when they felt that their rights had been violated. Here was a significant economic weapon against the cities in which universities were located. Such privileges made the university a powerful institution and enabled it to exercise considerable influence in medieval society.

By 1200 universities were flourishing in Bologna, Paris, and Oxford, having probably emerged in that order. Although documents that shed light on their origins and early development are scarce until the thirteenth century, by which time they were already well established, the emergence of universities was intimately associated with the new learning that had been translated into Latin during the course of the twelfth century. Indeed, the university was the institutional means by which Western Europe organized, absorbed, and expanded the great volume of new knowledge, the instrument through which it molded and disseminated a common intellectual heritage for generations to come. The earliest universities – Paris, Oxford and Bologna – were international in scope and easily the most famous of the Middle Ages. (Paris and Oxford were renowned as centers of philosophy and science; Bologna was equally notable for its schools of law and medicine.) By 1500 approximately seventy more universities had been created. Those in northern

Europe patterned themselves after Paris, whereas those in southern Europe chose Bologna as their model. From 1200 to 1500, three centuries of cultural and intellectual history shaped the university into a form that has persisted to the present day.

Although a detailed account of the structure and operation of the medieval university cannot be given here, a few indications of its organization will prove helpful. The medieval university was essentially an association of masters and scholars divided into four or fewer faculties (primarily arts, law, medicine, and theology) in each of which students matriculated toward either the baccalaureate or the master's degree. A master of arts degree was usually a prerequisite for entry into the higher faculties of law, medicine, and theology. Thus an arts master, teaching in the arts faculty, might also be a student matriculating for a bachelor's degree or a master's degree in theology, medicine, or law. The universities of Paris and Bologna provided two disparate models for the organization of universities founded during the remainder of the Middle Ages. Of the two models, only the University of Paris will be discussed here (despite its importance, the University of Bologna is much less relevant for natural philosophy).

The University of Paris was a "university of masters," so regarded because the masters of arts functioned as the governing body of the whole university. The arts masters of Paris controlled the curriculum, examinations, the admission of new masters, and the granting of the baccalaureate and master of arts degrees. The students and masters of the arts faculty – and only the arts faculty – were organized into four geographically based "nations," designated as French, Picard, Norman, and English (or English-German, which included students from central and northern Europe). Arts masters who subsequently became professors in the higher faculties of medicine, law, and theology retained full membership in their nations. The nations, each of which was headed by a proctor, actually governed the university, because they elected its head officer, the rector.

By modern standards, enrollments in medieval universities were small. The number of students at large institutions like Paris, Oxford, Bologna, and Toulouse was probably between one thousand and fifteen hundred. Approximately five hundred students entered the University of Paris every year. Because the average period of study for a student was approximately two years, the total number of students attending Paris at any given time was more than one thousand, perhaps around twelve hundred. During the course of the Middle Ages, however, the number of matriculating students appears to have increased. The long-range numbers are impressive. For the whole of Europe, scholars estimate that approximately seven hundred fifty thousand students matriculated at the universities between 1350 and 1500. The continually rising number of students also reflected the growing number of universities in that same period, when more than forty were founded. By the end of the Middle Ages, nearly every principal state in Europe included a university, founded either by a pope or a secular ruler. In retrospect, it is evident that no institution produced in the European Middle Ages has proven more permanent than the university.

## STUDENTS AND MASTERS

Most of the students at medieval universities departed after two years or less without acquiring a bachelor's degree. The percentage of students awarded that degree was thus relatively small. The longer the time required for the successful completion of a degree, the smaller was the percentage of students who received it. Whereas the bachelor's degree required three or four years, the master of arts degree demanded one or two years beyond that, for a total of five or six years of schooling. Occasionally, the time needed extended beyond, to seven or even eight years. The master of arts degree was a prerequisite for entry into one of the higher faculties of law, medicine, and theology, each of which demanded a number of additional years of study. Thus the number of students who successfully completed degrees in the higher faculties represented a small percentage of the total student community, perhaps smaller than the percentage of those who complete doctorates in modern universities. A student's attendance at a university, even for a short period of time, even without acquisition of a degree, was viewed favorably by society and was considered helpful in shaping the student's career.

During the Middle Ages, no hierarchy of educational institutions existed that was comparable to the sharp divisions between modern elementary schools, high schools, and colleges or universities. Previous attendance at a "lower-level" school was, therefore, not required for admission. Indeed, even the ability to read and write Latin may not have been essential. In the virtual absence of preconditions or prerequisites, entering a medieval university was relatively simple. There were, however, two indispensable requirements for admission.

The first was official matriculation, which was the responsibility of the university rector. To achieve matriculation, an entering student, who was usually fourteen or fifteen years of age, had to pay a fee and take an oath. The oath varied from university to university but usually involved the student's affirmation of loyalty to the rector and a promise to promote the welfare and integrity of the university. The student also swore that he would take no vengeance for any wrongs that might befall him. In return, the rector admitted the student into the university community and was thereafter expected to protect him when necessary. Despite its significance, the oath-taking ceremony itself was largely a formal exercise.

This was not true for the second indispensable requirement, however, which compelled each entering student to attach himself to a master. Students associated with the same master formed a natural group. Their academic fates were subject to his jurisdiction, and he was, accordingly, expected to introduce the students to the university community and to university life. He also prepared his students for examinations by seeing that they met the various test requirements in the proper sequence. Presumably, the master also laid out a course of study for his students whereby they attended his lectures over a period of three or four years and perhaps also took suggested classes offered by other masters. The selection of a master by a student was probably made on the basis of personal criteria, which might involve such considerations as geography, family connections, and friendships. It seems likely that master-student clusters allowed for more personalized relationships within the more formal and perhaps forbidding institutional structure of the university as a whole.

## TEACHING IN THE ARTS FACULTY

Teaching was the most important activity at medieval universities, but the teachers themselves – the masters – were relatively inconsequential. Although there were famous masters, their fame rarely depended on their teaching. Teachers were viewed as little more than interchangeable parts. At least two factors produced this state of affairs. The curriculum in medieval universities was much the same everywhere and, for the most part, was repeated each year. Because subject and area specialists did not exist in the arts faculties of medieval universities, elective courses formed no part of the curriculum. Every arts master was assumed capable of teaching any of the regular course offerings in natural philosophy (perhaps courses in the quadrivial subjects as well). In this sense, then, the masters were interchangeable.

The second factor, which reinforced the first, concerns teaching methods and techniques. University instruction centered upon the lecture (*lectio*) and the disputation (*disputatio*). Lectures were of two basic kinds, ordinary and extraordinary. Ordinary lectures lay at the heart of the teaching program and were always given in the morning by designated regent masters, that is, active teaching masters. As a sign of their importance, no other lectures or activities were permitted during the delivery of ordinary lectures. By contrast, extraordinary lectures were usually given in the afternoon, after ordinary lectures, or on a day when no ordinary lecture was scheduled. Extraordinary lectures were more flexible and informal than their ordinary counterparts and could be given by students as well as by masters. A third, less important, type of lecture, also offered in the afternoon, was called cursory and usually involved a summary or review of problems derived from a standard text.

The purpose of ordinary lectures was to present the required texts that comprised the official curriculum. Modern scholars have said little about what actually occurred in a typical medieval university classroom, probably because teachers and students left few descriptions of their classroom experiences. It is likely, however, that classroom lectures were a passive experience for the students, who simply listened and perhaps took a few notes. Students who had a copy of the text being discussed – and few did – might even follow along.

Lectures were primarily the province of the masters, who had considerable leeway to introduce their own opinions. In lectures that were at least an hour in length, and perhaps as long as two hours, a master of arts would take up some portion of a prescribed text, say Aristotle's *Physics* or *On the Heavens*. During the thirteenth century, a number of techniques were developed for presenting the text. Initially, the master read the official text and glossed terms and expressions that required explanation. Soon after, however, masters began to summarize the text and also add explanatory opinions and critical comments, integrating the two. The model for this approach may have been Avicenna's translated works. The Aristotelian commentaries by Albertus Magnus (Albert the Great) are a prominent example of the Avicennan technique.

Another method of presenting an ordinary lecture was to separate the text and the commentary. In this approach, the teacher, or commentator, did not merely explain each section of the text, but might also include the opinions of other commentators and authors as well as his own. Averroes's numerous commentaries on the works of Aristotle were of this kind and may have been the model for scholastic commentaries of the thirteenth century. Thomas Aquinas, Walter Burley, and Nicole Oresme were a few of the scholastics in the thirteenth and fourteenth centuries who followed the method of Averroes.

By the end of the thirteenth century another method of textual analysis had emerged that was destined to overshadow all others. Because medieval masters had a great degree of freedom in how they covered the required texts, some of them began to focus on special themes and problems that were inherent in the text, usually considering them near the end of the lecture. Gradually, however, the masters reduced the amount of straightforward, sequential commentary, replacing it with a discussison of special problems. In time, consideration of these special problems, or questions (*questiones*), completely replaced the commentary. The significance of the questions transcended the classroom, however, because the lectures of many teachers were written down and "published." Publication should be understood as a process whereby scribes at the university bookstore made master copies of the teachers' lectures. From a master copy, other copies could subsequently be made either for rent or for purchase by students and teachers. In this way, copies of a work were disseminated. What emerged was the most important category of scholastic literature, the questions format. This genre became almost synonymous with the notion of *scholastic method*, because, as we shall see, it utilized the basic form of a scholastic disputation.

Scholastic disputations, in which the students were prominent participants, were a vital aspect of university education. Whereas students may have been passive auditors in the medieval lecture halls, disputations offered them the opportunity to apply what they had absorbed. By analogy with lectures, disputations were divided into ordinary and extraordinary. The ordinary disputation (*disputatio ordinaria*) had the same exalted status as the ordinary lecture. Masters held these disputations on a regular basis, usually once a week, and required their students to attend. Other masters might also attend a colleague's ordinary disputation. The presiding master, however, posed the question, one that he might have wished to examine more carefully than his ordinary lectures permitted. The other masters and students participated, some supporting the affirmative side, others the negative side. It was the presiding master, however, who "determined" the question – that is, who synthesized the various arguments into a definitive answer to the problem.

In this exercise, students learned how to cope with contentious questions and thus gained valuable experience in preparation for their own ascension to mastership. For their first two years, students were usually silent observers. During their third and fourth years, however, they were expected to respond to questions and to offer answers. On the basis of this experience, and on the assumption that they met all the necessary prerequisites, successful student respondents were given permission to determine a disputation, that is, to supply the final answer to a question based on all the previous affirmative and negative arguments. With the successful completion of the determination (*determinatio*), the student became a bachelor of arts.

Bachelors of arts, who went on to study for the master of arts degree, had to spend at least two more years of study at the graduate level. In addition to attending lectures in natural philosophy, they usually spent time delivering afternoon lectures on texts assigned by their respective masters, either texts on logic, or, as was more common, on Aristotle's natural books. The bachelor was also expected to attend disputations by both masters and students. When this portion of a student's curriculum was completed to his master's satisfaction, the latter would recommend that his student be permitted to "incept" – that is, enter a two-stage process that terminated with conferral of the master of arts degree. In the first stage, the bachelor participated in a disputation in which he played the role of respondent to his master for the last time. During the second stage, the bachelor received the insignia of the mastership and then delivered a brief inaugural lecture, after which he presided over, and determined, two disputational questions.

As part of the requirements for the right to incept, the prospective new master had to swear that he would teach in the arts faculty for at least two years, giving ordinary lectures and presiding over weekly disputations. In addition to the "ordinary disputation," however, a master might from time to time undertake a "quodlibetal disputation" (*disputatio de quolibet*). Beginning in the theological faculty in the thirteenth century, and spreading to the arts faculty in the fourteenth century, masters held public disputations one or two times a year, usually at Advent and Lent. Because these were public disputations, anyone could attend: students, masters, and those who had no connection with the university but who wished to observe an unusual, intellectual free-for-all or who, for whatever reason, preferred to be indoors rather than outdoors during the time of the dispute.

In a quodlibetal disputation, one master presided. The complete disputation usually occurred over two days. The questions – and there would be many – were proposed by members of the audience. Any question was permissible, however controversial. Some questions were theologically and politically explosive, posed in the hope that they might embarrass the presiding master. But many questions – if not most – were on problems in natural philosophy. During the first day of the disputation, as many as thirty or forty disparate questions might be proposed. Members of the audience were eligible to participate. They could pose questions or respond to them. Tentative solutions to many questions were proposed. Because the questions were numerous, wide-ranging, and often unrelated, the master was not required to consider them in the order in which they had been proposed. Rather, he was expected to organize them into a manageable order before entering the public arena the following day, when he would demonstrate his virtuosity by "determining" – that is, definitively resolving – each question in the order in which he had arranged them. The quodlibetal dispute provided an emotional outlet for the university community, a release from the rigid format of the ordinary disputations and lectures.

## THE CURRICULUM OF THE ARTS FACULTY

To this point, we have seen the manner in which students acquired their degrees at the medieval universities and the teaching methods developed by their masters. It is now time to describe what masters taught and what students were expected to learn.

Prior to the introduction of Greco-Arabic science and natural philosophy, medieval "arts" education was based, as we saw in the first chapter, on the seven liberal arts. With the introduction of Aristotle's works, and of Greco-Arabic science in the late twelfth and thirteenth centuries, the primacy of the traditional seven liberal arts ceased, and they became pathways, or handmaidens, to philosophy, or more precisely, to natural philosophy. The new learning transformed the liberal arts. Three of the four subjects of the old quadrivium – arithmetic, geometry, and astronomy – were greatly enriched by Greco-Arabic science. The trivium of the seven liberal arts was also expanded, most significantly in the area of logic, or dialectic. Logic was the first of the seven liberal arts to be dramatically affected by the new learning, especially by Aristotle's "new logic," which consisted of treatises by Aristotle that had not been known in the West prior to the twelfth century (*Prior* and *Posterior Analytics*, the *Topics*, and *Sophistical Refutations*). Of the seven liberal arts, logic played the most significant role in the new curriculum, largely because it was perceived as a tool of analysis applicable to all fields, a role that Aristotle himself assigned to it when he called his logical works the *Organon*, or instrument. Apart from logic, however, which was part of the traditional trivium, the quadrivial subjects of the liberal arts receded somewhat into the background to be replaced at center stage by Aristotle's philosophy, which came to be subdivided into three parts, known collectively as "the three philosophies": natural, moral, and metaphysical. The curriculum of medieval universities was essentially comprised of logic, the quadrivial subjects, and the three philosophies, of which natural philosophy was clearly the most important.

### Logic

Logic was a technical subject that developed a terminology of its own to cope with a host of problems of language and inference. It was concerned with the properties of terms and how the context of a term affected its meaning, as well as with the relationships between propositions. Over the course of logic's medieval history, a large number of problems were encountered that required the invention of new terms and techniques. The very terms that came to be associated with this history testify to the richness of medieval logic and to the numerous concepts and techniques that were produced by its practitioners. By the sixteenth century, however, knowledge of medieval logic, with its complicated terminology, had almost vanished. As humanism became more significant in the fifteenth and, especially, the sixteenth centuries, humanist authors attacked what they regarded as the sterility and barbarity of medieval logic. Traditional terms and expressions, many based on Aristotle's *Topics*, were easy prey to their scathing criticisms. It became difficult to de-

fend a discipline with an array of terms such as "supposition," "signification," "univocation," "equivocation," "copulation," "appellation,"
"ampliation," "restriction," "categorematic," "syncategorematic," "consequences," "obligations," "exponibilia," "sophismata," and "insolubi-
lia." By the sixteenth century, humanist education placed emphasis on
the style and content of language, rather than on its formal aspects.
Moreover, medieval logic seems to have been taken as far as was possible
with a purely verbal form of expression. It needed to develop a formal
method of representation of the various possible logical relations in a
manner analogous to the development of symbolic algebra, which had
been progressing since the fifteenth century.

Although medieval logic was usually applied to hypothetical exercises
and problems, scholastic authors occasionally applied their knowledge
of formal logic to problems of natural philosophy, with the reasonable
assumption that their readers would understand its role in the discussion.

## The quadrivium

The quadrivium functioned as the source of theoretical and exact science
for medieval university students. It differed radically, however, from the
quadrivium in the curricula of the monastic and cathedral schools of the
early Middle Ages. The emphasis on the exact sciences in the universities
of the late Middle Ages was not of equal breadth and scope. At Oxford,
the exact sciences formed an integral part of the curriculum from the
thirteenth century onward, but they received much less emphasis at Paris
and elsewhere. At Paris, mathematics and the other quadrivial sciences
were rarely part of the regular course offerings. Mathematics, for example, was not usually taught at Paris in the thirteenth century and
taught there only sporadically in the fourteenth. Masters interested in
the exact sciences could offer such courses privately to interested students.

Although numerous works in arithmetic, geometry, astronomy, and
music, many of which had been translated from Arabic or Greek, were
extant in the Middle Ages, only a limited number were required texts in
university courses. Most treatises in the exact sciences were, however,
available for study. Indeed, many had been composed in the Middle
Ages by scholars trained at the university, where they first became con-
versant with the sciences. Of the four quadrivial sciences, arithmetic and
music bore some resemblance to their counterparts in the early Middle
Ages, whereas geometry and astronomy were virtually new sciences.
One author from the early Middle Ages, Boethius, provided the fundamental treatises in arithmetic and music, namely his *Arithmetica* and *Musica*. But in both of these subjects, treatises composed in the thirteenth

and fourteenth centuries advanced these disciplines far beyond Boethius. Although Boethius's *Musica*, along with Saint Augustine's treatise *On Music (De musica)*, were the standard texts in the teaching of music in arts courses, significant new treatises were composed in the fourteenth century by Johannes de Muris (John of Murs), Philippe de Vitry, and Guillaume de Machaut. These authors and others played a role in producing a musical notation. In arithmetic, Boethius's theoretical treatise was supplemented by books VII to IX of Euclid's *Elements*, which treated number theory, and by Jordanus de Nemore's (fl. ca. 1220) *Arithmetica* in ten books, which included more than four hundred propositions and became the standard source of theoretical arithmetic in the Middle Ages.

Geometry was the mainstay of the curriculum in the exact sciences, and Euclid's *Elements*, which was almost unknown during the early Middle Ages, was its fundamental text. Of the thirteen genuine and two spurious books of the medieval Latin version of the *Elements*, only the first six books were usually required. As did arithmetic, geometry had a practical, or applied, side. In the Middle Ages, its most important application was in astronomy. Among astronomical works, the best known and most important was Ptolemy's *Almagest*, which furnished the basis for technical knowledge of the subject.

Despite its appearance on curriculum lists, the *Almagest* was too technical for use as a text. Much simpler treatises were required. Two thirteenth-century works tried to meet this need. The most famous and most popular was John of Sacrobosco's *Treatise on the Sphere* (*Tractatus de sphaera*), the four chapters of which provided a brief survey of different parts of the finite, spherical universe. Although the fourth book was purportedly concerned with planetary motion, the subject's treatment was so meager that an unknown teacher of astronomy composed a work to remedy this deficiency. The *Theory of the Planets* (*Theorica planetarum*) introduced generations of students to the basic definitions and elements of planetary astronomy and provided them with a skeletal frame of the cosmos. On a more practical level, students were also taught something about calculating the various feast days in the ecclesiastical calendar. For this purpose, computational treatises were used under the generic title *compotus*, the most popular of which were probably those written by John of Sacrobosco and Robert Grosseteste. Geometry also played a role in determining the use of an astronomical instrument called the quadrant (for example, the *Treatise on the Quadrant* by Robertus Anglicus) and also found application in treatises on weights, or the science of statics, which were associated with the name of Jordanus de Nemore, and in treatises on perspective, or optics, in works associated with the names of Ptolemy, Alhazen (Ibn al-Haytham), John Pecham, and others.

The significance attached to the exact sciences in the university curriculum is not evident from curriculum lists, most of which did not survive and which are, in any event, spare of detail. We can best infer their importance from the attitudes of scholars, who were also university teachers. Geometry was no longer valued only for its practical use in measurement, or even as a vital aid for philosophical understanding. Roger Bacon and Alexander of Hales extolled its virtues as a tool for the comprehension of theological truth. They regarded geometry as essential for a proper understanding of the literal sense of numerous passages in Scripture, as, for example, those about Noah's ark and the temple of Solomon. Only by interpreting the literal sense with the aid of geometry could the higher, spiritual, sense be grasped. Geometry was also thought mandatory for a proper understanding of natural philosophy, as Robert Grosseteste argued in his treatise *On Lines, Angles, and Figures*. A universe that was constituted of lines, angles, and figures could not be properly interpreted without geometry; nor, indeed, could the behavior of light, which, like most physical effects, was multiplied and disseminated in nature geometrically.

Arithmetic was equally valued; indeed, it was often ranked first among the mathematical sciences. In his fourteenth-century treatise the *Commensurability or Incommensurability of the Celestial Motions*, Nicole Oresme provides insight into the way arithmetic was viewed and how its relationship to geometry might be perceived. Within the framework of an imaginary debate between geometry and arithmetic, Arithmetic presents herself as the firstborn of all the mathematical sciences and the source of all rational ratios and, therefore, the source of the commensurability of the celestial motions and the harmony of the spheres. Prediction of the future also relies upon exact astronomical tables, which depend for their precision on the numbers of Arithmetic. By way of rebuttal, Geometry claims greater dominion than Arithmetic, because she embraces both rational and irrational ratios. As for the beautiful harmony allegedly brought into the world by the rationality of Arithmetic, Geometry counters by noting that the rich diversity of the world could be generated only by a combination of rational and irrational ratios, which she alone produces.

Geometry and arithmetic were both valued because they were deemed essential for penetrating the operations of nature and for describing the variety of motions and actions in the world. The medieval emphasis on geometry and arithmetic should give pause to those who have argued that medieval natural philosophers and theologians were hostile to mathematics.

The science of astronomy, which included astrology, was also regularly praised as an essential instrument for understanding the universe. Astronomy could predict, but not determine, future events. Roger Bacon judged it essential for church and state, as well as for farmers, alchemists, and physicians; Robert Grosseteste regarded it as invaluable for many other sciences, including alchemy and botany. Music was also accorded high status. It was thought helpful in medicine because physicians could employ it as part of an overall regimen for health. Bacon also emphasized music as a factor in stirring the passions in war and soothing them in peace. Because musical expressions and instruments are frequently mentioned in Scripture, it was thought that the wise theologian would do well to learn as much about music as possible.

## The three philosophies

Although the seven liberal arts were augmented, and even transformed, in the late Middle Ages, they, nevertheless, represented the traditional format for education. The really new learning in the universities of the thirteenth century came with the introduction of Aristotle's philosophical works, which would form the major requirement for the degree of master of arts. Based on Aristotle's works, three major philosophical domains were distinguished: moral philosophy (or ethics), metaphysics, and natural philosophy. The primary text for the first of these subject areas was Aristotle's *Nicomachean Ethics*, while Aristotle's *Metaphysics* was obviously the text for the second. Of the three philosophies, Aristotelian natural philosophy was the most important and formed the core of a university education. Aristotle's *natural books* (*libri naturales*) served as the texts for the study of natural philosophy and included his *Physics* (*Physica*) and *On the Soul* (*De anima*), probably the two most important books in natural philosophy, along with his *On the Heavens* (*De caelo*), *On Generation and Corruption* (*De generatione et corruptione*), *Meteorology* (*Meteora*), and *The Small Works on Natural Things* (*Parva naturalia*). Although they were not usually the subject of lectures and were rarely, if ever, required texts, Aristotle's biological works also belong to the literature of medieval natural philosophy. In the Middle Ages, natural philosophy served as a foundation for moral philosophy and was almost everywhere interwoven with metaphysics; even theology drew heavily upon it, as did medicine and, occasionally, music. Because of its crucial importance, the focus in this volume will be on natural philosophy and how the problems with which it was concerned, and the methods used to resolve them, would ultimately prove invaluable for the development of early modern science.

## THE HIGHER FACULTIES OF THEOLOGY AND MEDICINE

Because they both made extensive use of natural philosophy, I must say something about the higher faculties of theology and medicine. Although theology schools did not usually require a master of arts degree for admission to their programs, most who entered had it, or had substantial training in the arts, especially in logic and natural philosophy. As we shall see in chapter 5, many theologians regarded logic and natural philosophy as essential tools for the elucidation of theological problems, even though ecclesiastical authorities often complained – even as late as the sixteenth century – that theologians were far too engrossed in these secular subjects for their own good and the good of theology.

With substantial backgrounds in natural philosophy, students were ready to engage in the lengthy study for a master (or doctor) of theology degree, a course of study that at different periods took anywhere from ten to sixteen years. Those who obtained the degree were often around thirty-five years old, quite an advanced age at a time when the average life span was probably no more than fifty years. Theology students studied two basic texts intensely: the Bible and the *Sentences* of Peter Lombard. In the lengthy course of study, a student heard lectures on the two basic texts during the first five to seven years, after which he became a "biblical bachelor" (*baccalarius biblicus*) and lectured on certain books of the Bible for two years. Those who advanced beyond this stage were ready to lecture for approximately two years on the *Sentences* and were therefore known as "Sententiary bachelors" (*baccalarii Sententiarii*). Upon completion of these lectures, the candidate became a "formed bachelor" (*baccalarius formatus*) for four more years during which he engaged in many of the activities of theological masters, such as, for example, giving sermons and conducting quodlibetal disputes. After these many years of study and training, the bachelor finally completed the requirements for the license to teach theology and for the degree of master of theology.

Among university disciplines, medicine was more intimately related to the arts than to theology. In preparation for the study and practice of medicine, astrology and natural philosophy played significant roles. Most of the students who studied in medical schools had either a master of arts degree or a reasonable background in the arts. It was fairly standard practice to reduce the length of study for those who were judged proficient in the arts. The length of study for a medical degree varied, but six to eight years was fairly common. As with the other faculties, students obtained their medical degrees by attending prescribed lectures on certain required texts, by engaging in disputations, and by taking oral examinations.

Because the overwhelming number of those who obtained the medical degree entered private practice, the medical curriculum was oriented toward practice, even though the texts were largely theoretical. Students were expected to acquire practical experience during summers by assisting physicians either at the university or in private practice. Beginning in the fourteenth century, they were also expected to attend dissections that were supposed to be held regularly.

The quantity of medical literature in the Middle Ages was large, and only selected texts could be used as the basis of lectures. Works translated from Arabic were fundamental and included numerous treatises by Galen (ca. 129–ca. 200), the great Greek physician, as well as the works of certain Muslim physicians, most notably Avicenna (Ibn Sina) (*Canon of Medicine*), Rhazes (al-Razi, d. 925) (the *Comprehensive Book*, or *Liber continens*), and Averroes (Ibn Rushd) (*Colliget*).

## THE SOCIAL AND INTELLECTUAL ROLE OF THE UNIVERSITY

The aims of the faculties of theology, medicine, and law are fairly obvious. They were professional schools. The purpose of a faculty of theology was to produce theologians; that of a faculty of medicine to produce physicians; and that of a faculty of law to produce lawyers. The texts that were studied in each of these faculties were chosen to facilitate its goals. But what was the objective of the arts faculty? What did bachelors and masters of arts aim to achieve with the curriculum I have just described? Of what value was an education based on logic, a few exact sciences, and natural philosophy?

The most obvious goal of the arts curriculum and the arts degree was to produce new teaching masters for the arts faculties of Europe. And, of course, some, if not many, arts masters earned their livelihoods as teachers. Indeed, new masters were required to teach for at least two years after receipt of their degrees. But what about masters who did not choose to make a career of teaching? What were the prospects for those students who had only a bachelor of arts degree or only a year or two of an arts education? Were there employment opportunities for individuals who had a few years of an arts education and were acquainted with logic, the quadrivium, and the three philosophies? For these individuals, the best opportunities for employment probably lay in a royal or ducal court, or in the Church, or perhaps even in a communal or municipal government. Even a relatively brief attendance at a university implied the ability to write Latin and at least a rudimentary knowledge of arithmetic calculations, which were useful skills for potential bureaucrats. But in many instances, former students may have been capable of drawing upon their overall education to contribute more than a bare minimum to prospective employers. After all, they had been exposed to many ideas about life and the physical world that were deemed important in their day.

And yet the arts curriculum I have described seems, at first glance, remote and irrelevant to the operations of medieval society. Why was it so heavily theoretical and devoid of practical courses that might have been more useful for the needs of society? Why did the medieval universities fail to include important practical subjects from the mechanical arts (*artes mechanicae*) like architecture, military science, metallurgy, and agriculture? Although the university community recognized the inherent value of the arts curriculum, and also acknowledged its value as a preliminary course of study for entry into the higher faculties of medicine, theology, and law, how society at large viewed an arts curriculum based on logic, bits and pieces of a few exact sciences, and an overwhelming dose of Aristotelian philosophy and natural philosophy is more difficult to ascertain.

In truth, the arts curriculum of the medieval university was not developed to meet the practical needs of society. It evolved from the Greco-Arabic intellectual legacy that came by way of the translations of the twelfth and thirteenth centuries. That legacy consisted of a body of theoretical works that were to be studied for their inherent value and not for practical reasons or monetary gain. The ancient tradition, exemplified by Aristotle and reinforced by Boethius and others, laid great emphasis on the love of learning, and on acquiring knowledge for its own sake. It scorned those who learned in order to earn a living or to make practical things. Teachers and students in medieval society fully subscribed to this viewpoint and shaped the medieval university accordingly.

Practicality, however, is in the eye of the beholder. The kind of theoretical learning that was emphasized in antiquity and the Middle Ages (see chapter 7) may have been perceived as eminently pragmatic and sensible. From such learning one could derive knowledge about the way the world functioned, and thereby acquire precious insight into the perpetual causes and effects that shaped human existence. Many would have regarded such knowledge as more worthy than any other kind and therefore as eminently practical. Whatever their ultimate attitude, medieval scholars thought it important to know about the structure and operation of the universe, which was what an arts education was all about.

With the acceptance of the universities by church and state, society as a whole came to accept the university arts ideal of learning, an ideal that was assumed to be of great personal value to the individual but of little direct value to the mundane activities of society. This state of affairs persisted for centuries. No significant expansion of the arts curriculum occurred during the Middle Ages. Not until the Renaissance did changes occur, and even then, the expansion inclined toward the inclusion of humanistic subjects, such as history and poetry, which were lacking during the Middle Ages, rather than in the direction of practical subjects. Indeed, the ideal of ancient and medieval learning – to acquire knowledge for its own sake – remained largely intact.

If the arts programs at medieval universities failed to provide practical benefits to society, they nevertheless did construct firm foundations for the development of science and the scientific outlook. This occurred only because of the unusual structure and traditions of the university, that unique institutional contribution of the Middle Ages to Western civilization. Its extraordinary accomplishments even filtered through to the Arab world. "We further hear now," declared the great Islamic historian Ibn Khaldun (1332–1406),

> that the philosophical sciences are greatly cultivated in the land of Rome and along the adjacent northern shore of the country of the European Christians. They are said to be studied there again and to be taught in numerous classes. Existing systematic expositions of them are said to be comprehensive, the people who know them numerous, and the students of them very many.[1]

Although the medieval university was radically different from any institution known to the ancient Greeks, Romans, and Arabs, it is familiar to the students and faculty of any modern university, which is, after all, its direct descendant.

## THE MANUSCRIPT CULTURE OF THE MIDDLE AGES

Before the advent of printing in the middle of the fifteenth century, treatises in medieval science and natural philosophy depended for their existence on manuscript copies. As a consequence, they were subject to all the vagaries and uncertainties of any system that must rely on a scribe or copyist to produce one or more copies from an exemplar or to record a lecture as it was given. Medieval Latin texts were subject to more than the ordinary scribal vicissitudes – errors of commission and omission – because medieval copyists had developed an elaborate system of abbreviations that served to speed the process of copying and also tended to save paper. These abbreviations frequently added an element of uncertainty to an interpretation of the text, both for someone who wished to read it as well as for someone who wished to copy it. The difficulties in deciphering medieval manuscripts affect modern understanding of medieval science in two basic ways.

The first way in which the difficulties in deciphering medieval manuscripts affects our understanding today concerns the integrity of an author's work as it was copied and recopied and read by students and scholars over the course of centuries. Because copies might vary drastically as a result of scribal errors introduced at any point in the dissemination process, we may infer that the reader's understanding of an author's intent in some, and perhaps in many, passages was almost unavoidably distorted. Reliance on handwritten and handcopied works meant that versions of the same treatise in Paris, Oxford, and Vienna might differ substantially. In astronomical and mathematical texts, for example, essential diagrams and figures may have been included in some versions, but omitted or included only partially in others. Even when a diagram was included, scribal errors might reduce or destroy its utility. In purely verbal texts, words might be omitted or added by the scribe. Many of the copies of medieval works that have survived were not made by professional scribes, but by students who had copied the texts for their own use. Such copies were often passed along to other students, who would introduce more errors and changes. To these formidable problems, we must add that of legibility. The handwriting of copyists was frequently difficult to decipher and all too often was simply unintelligible.

University stationers, or booksellers, had as their responsibility the production of reliable texts for university personnel. They would often receive the pristine version of a treatise directly from its author. From this original they would make one or more copies. The stationers were authorized to lend all or parts of the texts to students who, for a fee, could copy it for their own use. Obviously, student copies varied in quality. Many were subsequently passed on to other students for further copying. Errors were inserted at virtually every stage of the process of multiplying and disseminating texts. Perhaps the only exception to this generalization are copies of the Bible, which were carefully supervised.

The second way that the interpretation of medieval manuscripts may affect our understanding of medieval science has to do with the limits imposed on modern scholars who read or edit treatises written in the Middle Ages. Most scholars would probably begin with a list of the extant manuscripts of the treatise in question. The quality of those manuscripts, which managed to survive the ravages of time, determines the level of intelligibility of that treatise. In most instances, significant gaps in our understanding of that treatise will probably remain even after modern scholars have completed an edition of it.

It is evident that differences between an original version of a medieval treatise and all the copies that were subsequently made from it were at best considerable and at worst vast. From our vantage point, we can see how difficult it must have been to do science in the Middle Ages. The preservation of reasonably faithful versions of the basic Greco-Arabic texts that had been translated into Latin was itself a major task. To this we must add the vast array of medieval scientific texts, commentaries, and questions that were copied and recopied. Unfortunately, not all texts were copied and recopied. Many treatises simply disappeared. During the Middle Ages, knowledge was as likely to vanish as to be preserved. An enormous effort would have been required just to maintain the status quo, or to restore a text that had been corrupted. Although we cannot measure the detrimental effects on medieval science and natural philosophy that followed solely because of a dependence on handwritten manuscripts, we may plausibly conjecture that they were enormous.

The introduction of printing in the mid-fifteenth century significantly altered this picture. With the advent of printed books, knowledge in general, and technical information in particular, could be disseminated with a speed and accuracy that could scarcely have been imagined in the age of manuscripts. Science was a particular beneficiary of printing. Identical copies of a scientific work could be spread through Europe in a relatively brief time. And yet, the precise role of printing in the generation of the Scientific Revolution is in dispute. We must ask if, in the absence of printing, the old scribal system could have been improved to multiply copies of scientific treatises and thereby meet Europe's intellectual needs. And would the ever-expanding royal, ducal, municipal, and university libraries have provided European scholars with sufficient access to allow the continued expansion of science and learning? Fortunately, we need not answer these questions in this study. The foundational contributions to early modern science that are its focus had already been formed long before Gutenberg's printing press converted Europe from a manuscript to a print culture.

Although manuscript reproduction and dissemination posed serious problems in the Middle Ages, we must not conclude that the problems were insuperable. Despite the formidable obstacles just described, the quality of the handwritten texts available to medieval scholars in science and natural philosophy was often more than adequate to allow for their comprehension and for the addition of significant contributions to learning. The legacy that has reached us is one that we can comprehend and often admire. The core of that legacy was Aristotle's natural philosophy, which was deeply rooted in the medieval university, and which I shall now briefly describe.

# 4

# *What the Middle Ages inherited from Aristotle*

ARISTOTLE'S natural books formed the basis of natural philosophy in the universities, and the way in which medieval scholars understood the structure and operation of the cosmos must be sought in those books. By his use of assumptions, demonstrated principles, and seemingly self-evident principles, Aristotle imposed a strong sense of order and coherence on an otherwise bewildering world. Aristotle's medieval disciples, who formed the class of natural philosophers during the late Middle Ages, would eventually extend Aristotle's principles to activities and problems beyond anything that the philosopher himself had considered.

Aristotle was convinced that the world he sought to understand was eternal, without beginning or end. He regarded the eternity of the world as far less problematic than any assumption of a cosmic beginning that also implied a future end to the world. It was better to postulate eternity than be forced into an explanation that required an infinite regress of causal beginnings. The idea that matter could have a beginning seemed impossible to the ancient Greeks, for if one were to arrive at some alleged pristine matter, it would inevitably lead to the question of what caused it, and so on. Without a beginning, however, the world could not have been created, and thus Aristotle's ideas about the eternity of the world set him in opposition to the theologians of the great monotheistic religions of Judaism, Christianity, and Islam. Of all the issues that involved natural philosophy and theology during the thirteenth century in Western Europe, theologians regarded the eternity of the world as the most difficult and threatening for the faith (see chapter 5).

Still, if Aristotle's world was eternal and therefore suspect, his insistence on its uniqueness placed him squarely in agreement with the sacred scriptures of the three great religions. He regarded our world as unique, a large finite sphere beyond which nothing could exist. All existent matter is contained in our world, with none left over. Without body, "neither place, nor void, nor time" could exist beyond the world, because the definitions of "place," "void," and "time" all depended on the existence of body. For Aristotle the proper place of a body was always the innermost surface of another immediately surrounding body that was in direct contact with the contained body. Thus a place is defined as something in which body must be present. Without the existence of a body beyond our world, no place could exist (for more on place, see later in this chapter). Similarly, a void is something in which the existence of a body is possible, though not actual. Therefore, if no body is possible, no vacuum is possible. Finally, time is the measure of motion. Without body there can be no motion and, therefore, no time. Aristotle concluded that all existence lay within our cosmos, and nothing beyond. The "nothing" in this sense is not to be construed as a vacuum, but is best characterized as a total privation of being.

Perhaps the most momentous decision that Aristotle made about the eternal, physical world was to divide it into two radically different parts, the terrestrial, which extended from the center of the earth to the lunar sphere, and the celestial, which embraced everything from the moon to the fixed stars. In the terrestrial region, observation and experience made it obvious that change was incessant, whereas in the celestial region change was virtually nonexistent. Astronomical observations inherited from the past convinced Aristotle that no changes had ever been detected in the heavens (*De caelo* 1.3.270b.13–17), from which he inferred that changes did not – and could not – occur there. To understand Aristotle's world better, it is advantageous to describe first the terrestrial region of change, which, in turn, will make the unchanging properties and attributes of the celestial region more comprehensible.

## THE TERRESTRIAL REGION: REALM OF INCESSANT CHANGE

Much of Aristotle's natural philosophy is an attempt to identify and explicate the principles of change in the terrestrial region, principles that shaped medieval interpretations of the processes that make the world what it is. Although we live in a world that had no beginning, Aristotle nonetheless explains how the development of matter is to be imagined and how it is differentiated into four basic elements – earth, water, air, and fire – that form the building blocks of all material bodies in the terrestrial region. The underlying basis of all material bodies is prime matter, which, although real, has no independent existence. Aristotle simply infers its reality because it was essential to assume the existence of some kind of substratum in which qualities and forms could inhere and produce sensible matter. Prime matter has no properties of its own, but is always associated with qualities that inhere in it and define it.

Which properties or qualities would raise prime matter to a higher existent level, say to the level of an element? After eliminating a number of possibilities, Aristotle argues that two pairs of contrary, or opposite, qualities could achieve this effect: hot and cold, and dry and moist. Because nothing could be simultaneously hot and cold, or dry and moist, no single pair of opposite qualities could inhere in prime matter at the same time. Non-opposite combinations, however, are possible and can produce elements. If the qualities coldness and dryness inhered in prime matter, they would produce the element earth; coldness and wetness would produce water; hotness and wetness air; and hotness and dryness fire. Thus were the four elements derived. The perceptible bodies of the terrestrial region were, however, not pure elements, but mixtures, or compounds, of two or more of them, usually called "mixed" bodies in the Middle Ages.

In Aristotle's natural philosophy, or physics, every body is a composite of matter and form, where the matter serves as a substratum in which the form inheres. The form of a thing, or a body, is its essential defining characteristics, the properties that make it what it is. Nature in the terrestrial realm is nothing more than a collective term for the totality of existent bodies, each comprised of matter and form. Every such body belongs to its own species and possesses the properties and characteristics – that is, the form – of its species. If unimpeded, it will act in conformity with those properties. Aristotle thus attributed to the bodies of the world a power to act in accordance with their natural capabilities. In this way, he allowed for secondary causation, where bodies were capable of acting on other bodies, that is, able to cause effects in other bodies. Aristotle believed that each effect was produced by four causes acting simultaneously, namely a material cause, or the thing out of which something is made; a formal cause, or the basic structure to be imposed on something; an efficient cause, or the agent of an action; and the final cause, or the purpose for which the action is undertaken. The causes that produce a stone not only make it heavy, but, if the stone is otherwise unimpeded, that heaviness confers upon it the capacity to fall naturally toward the center of the earth with a rectilinear motion. Similarly, the agents that produce fire confer lightness upon it and therefore the capability of rising naturally upward, whenever it is unhindered.

Aristotle was also concerned about the kinds of changes that the four causes could produce, distinguishing four kinds: (1) substantial change, where one form supplants another in the underlying matter, as when fire reduces a log to ash; (2) qualitative change, as when the color of a leaf is altered from green to brown in the same underlying matter; (3) change of quantity, as when a body grows or diminishes while otherwise retaining its identity; and, finally, (4) change of place, when a body suffers change as it moves from one place to another.

Of these four types of change, only the first and fourth require explanation. Substantial change is the most basic form of change, involving generation and corruption. For Aristotle every substantial change implied that something had come into existence from the passing away of something else. This coming-to-be and passing-away of things was the basis of all change in the terrestrial region. It occurred in all substances composed of matter and form, which in the terrestrial region included all things. Forms, or qualities, were potentially replaceable by other forms that were their contraries. When this occurred, one substance was changed into another. For example, fire, which possesses the primary qualities of hotness and dryness, is changed into earth, which possesses the primary qualities of dryness and coldness, when the hotness in fire is replaced by coldness, its contrary quality, or form. While one form is actualized in matter, its contrary is said to be in privation but potentially capable of replacing it. Eventually, each potential form or quality must actually become what it is capable of becoming; otherwise a form would remain unactualized, and nature would have produced it in vain. While one of a pair of contrary forms is actualized in matter, its contrary is absent and in privation, because two contrary forms cannot exist simultaneously in the same body. Virtually all change, that is, generation and corruption, involves the possession of one, and the exclusion of another, of a pair of contrary forms or qualities.

The last of the four changes, change of place, represents what we ordinarily think of as motion, the removal of a body from one location to another. Aristotle's doctrine of place may be viewed in two ways. In its broadest signification, it concerns the structure of the sublunar world; and in the narrowest and most restrictive sense, it involves the specific place of a single body. The broad sense of place is really the doctrine of natural place, in which Aristotle conceived of the part of the world below the moon as a structured region divided into four concentric areas, each the natural place of an element, toward which that element would naturally move if unimpeded. Thus the outermost concentric ring, located just below the concave surface of the lunar sphere, is the natural place of fire; the next concentric ring is the place of air, toward which air rises if in the regions below, or toward which it would fall if, for some reason, it was located in the region of fire; below air is the ring of water; and below that the sphere of our earth, the center of which coincides with the geometric center of the universe.

The earth's sphericity was a basic truth of Aristotle's system of the world. As observational evidence of its sphericity, Aristotle pointed to the curved lines on the Moon's surface during a lunar eclipse, inferring rightly that these were cast by the shadow of a spherical earth interposed between the Sun and the Moon. He also noted that as one changed position on the earth's surface, different stellar configurations came into view, indicating that the earth possessed a spherical surface. The sphericity of the earth seemed further confirmed by the way bodies were observed to fall to the earth's surface in nonparallel lines that met at its center. If all earthy bodies fell in this manner, they would cluster around the center of the world and form naturally into a sphere. So reasonable were Aristotle's arguments that a spherical earth was readily accepted.

What, however, about the place of any particular body? Aristotle's doctrine of place is based upon a fundamental conviction that the world is a material plenum in which the existence of void space is impossible. From this it followed that the place of any individual thing in the sublunar region consisted of the matter that surrounded it; or, as Aristotle described it, the place of a thing is "the boundary of the containing body at which it is in contact with the contained body."[1] The boundary, or innermost surface of the container, was also required to be motionless, a qualification that posed serious problems in the history of Aristotle's doctrine of place. It frequently happened that where the condition of contact was met, that of immobility was not, and vice versa. Nevertheless, when a body met these stringent conditions, it was presumed to be in its "proper place," that is, in a place that it alone occupied. Places that included more than one distinct body were characterized as "common places." Because Aristotle assumed that every body was somewhere, and therefore necessarily in a place, he was inevitably led to ask whether the outermost surface of the outermost sphere that contained the world was itself in a place, a question tantamount to asking whether the world itself is in a place. Convinced that bodies did not exist beyond the world, Aristotle argued that if no material body, and therefore no surface of a body, could surround our world, no body could function as its place. Paradoxically, although every body in the world is in a place, the last sphere, or the world itself, is not directly in a place. Apparently uneasy with this consequence of his doctrine of place, and perhaps fearful of being perceived as inconsistent, Aristotle found a kind of place for the last sphere by arguing that the last sphere is in a place indirectly by means of its parts, because "on the orb one part contains another."[2] Many of Aristotle's commentators rejected his cryptic attempt to find a place for the last sphere. And those who did not were often led into bizarre explanations to defend the master, as when Averroes argued that the last sphere is in a place accidentally (*per accidens*) because its center, the earth, is in a place essentially (*per se*). Thomas Aquinas thought it "ridiculous to say that the last sphere is in place accidentally [simply] because the center is in a place."[3] How could a container be in place by virtue of the thing it contains?

## Motion in Aristotle's physics

The motion of bodies from place to place was a problem that Aristotle frequently considered, although nowhere in his extant works is there a systematic and comprehensive treatment of it. The account that follows is based upon discussions scattered through a number of his works, especially the *Physics* and *On the Heavens*.

In a sublunar world that had no empty spaces and was a material plenum, motion, or local motion, as it was sometimes called, had to be from one place in that plenum to another. Aristotle distinguished two kinds of motion: natural and violent (or unnatural), a division that probably originated in gross observation. The division of local motion into natural and violent and the cluster of concepts, arguments, and physical assumptions associated with these two contrary motions formed the core of Aristotle's sublunar physics.

*Natural motion of sublunar bodies.* Aristotle's concept of natural motion was dependent on obvious properties he observed in the four elements – earth, water, air, and fire – that formed the material basis of all terrestrial bodies. When falling from heights, some bodies, like stones, were seen to move in straight lines toward the center of the earth. Other bodies, such as fire or smoke, always seemed to rise toward the lunar sphere and away from the earth's center. Because the class of bodies that fell naturally toward the center of the earth was, on the basis of experience, observed to be heavier than the classes of bodies that rose, Aristotle concluded that, when unimpeded, a heavy, or earthy, body moved naturally downward in a straight line toward the center of the earth. Thus the center of the earth – or more precisely, the geometric center of the universe – was the natural place of all heavy bodies. Conversely, light bodies moved naturally upward in a straight line toward the lunar sphere, which was conceived as their natural place. Aristotle described these natural up-and-down motions as accelerated.

Let us now apply these generalizations specifically to the four elements. Whenever an elemental body, composed primarily of earth, was above its own natural place – whether that place was in water, air, or the fiery region above the air – it was deemed absolutely heavy because, if unimpeded, it would fall toward toward the earth's center. Fire was regarded as absolutely light; if unimpeded, fire would always rise upward from the regions below toward its natural place above air, and below the lunar sphere. To emphasize fire's absolute lightness, Aristotle declared it "a palpable fact" that "the greater the quantity [of fire], the lighter the mass is and the quicker its upward movement."[4] By assuming that the greater the quantity of fire, the lighter it becomes and the faster it rises, Aristotle seems to have dissociated absolute lightness from the concept of weight, a concept that is unintelligible in this context. As for water and air, Aristotle regarded them as intermediate elements possessing only relative heaviness and lightness. When below its natural place somewhere within the earth, water would naturally rise; but when above its natural place, in air or fire, it would fall. Air, however, would fall when in the natural place of fire, but rise when in the natural place of earth or water.

Thus far we have described the idealized, natural behavior of each of the four elements. But the elements did not exist in a naturally pristine state. In the real world, bodies were actually compounds made up of varying proportions of all four elements. Bodies that fell naturally toward the earth's center did so because their predominant element was heavy (the heavier the body, the greater its speed of descent); those that rose naturally upward did so because they were dominated by a light element (the greater the quantity of air or fire in an airy or fiery body, the greater would be its speed of ascent).

Three pairs of opposites played a significant role in Aristotle's interpretation of the structure of the terrestrial, or sublunar, world. They may be schematized as follows:

1. Concave surface of lunar sphere        Geometric center of universe (or center of the earth)
2. Up                                      Down
3. Absolute lightness (fire)               Absolute heaviness (earth)

These opposites served as virtual boundary conditions for Aristotle's scattered account of the motion of bodies. The left-hand column tells us that an absolutely light body (fire) would naturally rise rectilinearly upward toward the lunar sphere, whereas the right-hand column informs us that an absolutely heavy body would fall naturally downward in a straight line toward the earth's center. Although Aristotle knew that earth was denser than air and water, he would have denied that density explained the fall of a stone through air or water. A stone falls only because it is absolutely heavy. Fire does not rise to its natural place near the surface of the lunar sphere because it is less dense than earth, water, or air, but rather because it is absolutely light. Indeed, fire does not even possess weight in its own natural place, so that if the air below were removed fire would not fall or move downward. With hindsight, we can now see that Aristotle's introduction of absolute heaviness and lightness was hardly conducive to the advance of physics, though Aristotle himself viewed it as a significant improvement over Plato and the atomists, who had attributed weight to all things, and for whom weight was a relative concept. Of the two possibilities before him, Aristotle chose the option that would prove historically least helpful. He did so, however, because he made his system heavily dependent on a variety of absolute contraries, choosing to avoid the relativistic comparisons of Plato and the atomists.

To provide a causal explanation for natural motion (and, as we shall see, for violent, or unnatural, motion as well), Aristotle invoked the general principle that every effect has a cause and assumes that every animate and inanimate thing capable of motion is moved by something else that is itself in motion or at rest.[5] (Or, to use the succinct medieval version of this principle: "everything that is moved is moved by another.") The mover and the thing moved were always assumed to be distinct entities. Although it might appear that natural motions would not require causal explanations because they are "natural," Aristotle assigned a particular agent (called the *generans*, or generator, in the Middle Ages) as the primary cause of unobstructed natural motion. The causative agent, or generator, was the thing that had originally produced the body actually in motion. For example, a fire produces another fire (as when a log is set ablaze) and confers on the new fire all the properties that belong to fire, one of which is the spontaneous ability to rise naturally when unimpeded. Similarly, whatever natural agent produces a stone confers upon it all of its essential properties, including the natural tendency to fall to earth when removed from its natural place.

Although he identified the *generans*, or the generator of a thing, as a kind of remote motive cause in natural motion, Aristotle interpreted the fall of a body as if its weight were the immediate cause of its natural downward motion; and he treated the rise of a body as if its lightness were the immediate cause of its natural upward motion. All other things being equal, Aristotle concluded that velocity is directly proportional to the weight of the body in natural motion and inversely proportional to the resistance it meets, which is measured by the density of the medium through which it moves, and that the time of its motion is directly proportional to the resistance, or density, of the medium and inversely proportional to its weight. For example, the speed of a body could be doubled either by doubling its weight (but holding the medium constant) or halving the density of the medium (and holding the weight of the body constant). Similarly, the time of motion could be doubled either by doubling the density of the medium (but keeping the weight constant) or by halving the weight of the body (and holding the density of the medium constant). Despite his recognition that unimpeded heavy bodies accelerated as they approached their natural places, Aristotle discussed natural motions as if their speeds were uniform.

*Violent, or unnatural, motion.* Motions that are violent, or unnatural, occur when bodies are pushed out of, or away from, their natural places. Thus a stone that is thrown rectilinearly upward into the air, or is hurled with a horizontal trajectory, is in violent motion; the motion of a fire that is somehow forced downward out of its natural place toward the earth is unnatural, or violent. Similarly, the motion of air when it is forced out of its natural place down toward the earth or upward toward the natural place of fire is characterized as a violent motion. Aristotle formulated specific rules in which he described the consequences that would follow from the application of a motive force to a resisting object. Although the rules are couched in terms of force, resisting body, distance traversed, and time, rather than directly in terms of velocity, the latter permits a more convenient summary. The velocity of a body in violent motion is inversely proportional to its own resistive power, which is left undefined, and directly proportional to the motive power, or applied force. In symbols, $V \propto F/R$, where $V$ is velocity, $F$ is motive force, and $R$ is the total resistance offered to the applied force, a quantity that, presumably, includes the resisting object or body plus the resistance of the external medium in which the motion occurs. To double a velocity, $V$, the resistance $R$ could be halved and $F$ held constant; or, $F$ doubled and $R$ held constant. To halve $V$, $F$ could be halved and $R$ held constant; or $R$ doubled and $F$ held constant.

Violent motion required a radically different causal explanation than did natural motion. The initial mover, or causal agent, was readily identifiable because it had to be in direct physical contact with the body it moved. Someone throwing a stone upward or pushing a wagon along a road is the mover, or motive power, in those violent motions. But the source of power that enabled a body to continue its motion after losing contact with its initial mover was far from obvious. How, for example, did a stone continue its motion after losing contact with the hand of a child that threw it? Aristotle argued that the external medium – air in the example of a stone – was the source of continuous movement. He believed that the original mover not only puts the stone in motion but also activates the air simultaneously. Apparently, the first portion or unit of activated air pushes the stone and simultaneously activates the adjacent, or second, unit of air which moves the stone a bit further. The second unit, in turn, simultaneously activates the next, or third, unit of air, and so on. As the process continues, the motive power of the successive units of air gradually diminishes until a unit of air is reached that is only capable of activating the very next unit of air, but is unable to communicate to it the power to move the body further. At that point, the stone begins to fall with its natural downward motion. By this mechanism, Aristotle employed the medium as both motive power and resistance. Not only did he believe that the medium as motive force had to be in constant physical contact with the body it moved, but he was equally convinced that the same medium had to function as a brake on the motion of that same body in order to prevent the impossible: the occurrence of an infinite speed, or an instantaneous motion. Aristotle took it as obvious that resistance to motion increased as the density of the medium increased, and decreased as the medium was rarefied. Because an indefinite rarefaction of the medium would result in a proportionate and indefinite increase in speed, Aristotle concluded that if a medium vanished entirely, leaving a vacuum, motion would be instantaneous (or beyond any ratio, as he put it).

The absurdity of an infinite speed was only one of a number of arguments that prompted Aristotle to reject the existence of a vacuum. The fundamental principles that he believed operative in the world would be useless in void spaces. Motion would be impossible for a number of reasons. The homogeneous nature of an extended void space meant that every part must be identical to every other part. Because differentiable natural places could not exist in a homogeneous space, bodies would have no good reason for moving in one direction rather than another. Natural motions would be impossible, as would violent motions, because the external medium Aristotle deemed essential for violent motion would be absent. If the void were infinite, and motion could somehow occur, that motion either would be unending – for what would stop a body in motion in a void that lacked other bodies and natural places to bring it to rest – or, in the absence of external resistances, it would be instantaneous. Among Aristotle's remaining arguments against the void, one is noteworthy. Bodies of different weights would necessarily fall in a void with equal velocities, which Aristotle regarded as an absurdity, because they ought to fall with speeds that are directly proportional to their respective weights. But the latter relationship could only occur in a plenum, where a heavier body cleaves through the material medium more easily than does a less heavy body. In the absence of a medium, Aristotle saw no plausible reason why one body should move with a greater speed than another. He therefore concluded that the world was necessarily a plenum, filled everywhere with matter.

## THE CELESTIAL REGION: INCORRUPTIBLE AND CHANGELESS

The part of the world that Aristotle envisioned beyond the convex surface of the sphere of fire was radically different from the terrestrial part just described. Aristotle regarded the celestial region as so incomparably superior to the terrestrial that he assigned to it properties that emphasized these profound differences. If incessant change was basic to the terrestrial region, then lack of change had to characterize the celestial region. This conviction was reinforced for Aristotle by his belief that human records revealed no changes in the heavens. Because the four elements of the sublunar region were involved in ceaseless change, they were obviously unsuitable for the changeless heavens. In his *On the Heavens* (bk. 1, chs. 2 and 3), Aristotle contrasted the natural rectilinear motion of the four sublunar elements (earth, water, air, and fire) with the observed regular, and seemingly natural, circular motion of the planets and fixed stars in the celestial region. The contrast between the straight line and the circle, the former finite and incomplete, the latter closed and complete in itself, convinced Aristotle, if he needed convincing, that the circular figure was necessarily and naturally prior to the rectilinear figure. Because the four simple elemental bodies moved with natural rectilinear (upward and downward) motion, Aristotle concluded that the observed circular motion of the celestial bodies must necessarily be associated with a different kind of simple, elemental body: a fifth element, or ether.

As if to emphasize the special importance of the ether, Aristotle often called it the "first body." Its primary properties were almost the opposite of those of the terrestrial elements. Where terrestrial elements moved naturally with rectilinear motions, the ether moved naturally with circular motion, which was superior because the circle was complete in itself, whereas the straight line was not. Where the four elements and the bodies compounded of them were in a continual state of flux, the celestial ether suffered no substantial, qualitative, or quantitative changes. Substantial change was impossible because Aristotle assumed that the pairs of opposite, or contrary, qualities, such as hotness and coldness, wetness and dryness, rare and dense, which were basic forces for change in the terrestrial region, were absent from the heavens and therefore played no role there. Aristotle's rejection of contrary qualities in the heavens led him to deny the existence there of the contrary qualities lightness and heaviness, from which he concluded that the celestial ether could be neither light nor heavy. Lightness and heaviness in the terrestrial region were associated with upward and downward rectilinear motions: heavy bodies approached the earth when they moved naturally downward, and light bodies receded from the earth when they moved naturally upward. In the absence of heaviness and lightness in the heavenly region, Aristotle inferred that rectilinear motions could not occur there. Thus not only was it observationally evident that the celestial motions were circular, but, from the very properties of the ether itself, it was apparent to Aristotle that rectilinear motions were impossible in the celestial region.

Because planets and stars are observed to move around the sky, Aristotle inferred that change of position was the only kind of change possible in the heavens. Celestial bodies continually change their positions by moving around the sky with effortless, uniform, circular motion. This uniform, circular motion is a natural motion, just as rectilinear up-and-down motions are natural. But where up and down were contrary terrestrial motions, circular motion had no contrary. Aristotle concluded that circular motion, which lacked a contrary motion, was natural to bodies composed of celestial ether, which lacked contrary qualities. In the absence of all contraries, change as it was observed in the terrestrial region could not occur in the ethereal heavens. Celestial bodies had to move eternally around the heavens with natural, uniform, circular motion. Although they changed positions, the absence of contraries prevented variations in their distances. Aristotle thus assumed that celestial bodies neither approached the earth, nor receded from it.

Aristotle associated change with matter, but he denied change in the heavens. Did it follow then that the heavens lacked matter and that the celestial ether, whatever else it might be, was not to be thought of as matter? On this important issue, Aristotle's remarks are inconclusive, and medieval natural philosophers were left to ponder his meaning. Both interpretations – that matter exists and does not exist in the heavens – received support.

Whether or not it was to be construed as matter, the celestial ether posed other problems. Because it was a perfect substance extending from the moon to the fixed stars, Aristotle seems to have thought of the ether as homogeneous, with all its parts identical. A glance at the heavens should have dispelled such a notion. At the very least, the celestial region consisted of visible bodies surrounded by empty portions of sky, a configuration that hardly suggests homogeneity. If celestial bodies and empty sky were both composed of the same ether, why did they differ? Why were planets and stars visible, and the rest of the sky effectively invisible? If the planets were made of the same ether, why did they seem to differ from one another? Why did their properties vary? To these questions, Aristotle supplied no answers, perhaps because the questions never occurred to him. When such questions occurred to his Greek, Arabic, and Latin commentators, they had to devise their own responses, a common fate for those who spent much of their lives seeking the meanings of Aristotle's texts.

On the nature of the empty celestial spaces, however, Aristotle was quite clear: they were filled with invisible, transparent, ethereal spheres that were nested one with another and each of which turned with regular, uniform, motion. Celestial bodies – planets and fixed stars – were somehow embedded in these spheres and carried around by them. Aristotle based his system upon the earlier mathematical systems of concentric spheres devised by Eudoxus of Cnidus and Callippus of Cyzicus in the fourth century B.C. In the latter's scheme, on which Aristotle directly founded his cosmology of concentric spheres, the planet Saturn, for example, was assigned a total of four spheres that were supposed to account for its celestial position. Of these, one was for Saturn's daily motion; one was for its proper motion along the zodiac, or ecliptic; and two represented its observed retrograde motions along the zodiac. Aristotle transformed Callippus's mathematical spheres into a system of real, earth-centered, physical celestial orbs that were collectively coterminous with the celestial region. To prevent the transmission of Saturn's zodiacal and retrograde motions to Jupiter, the planet immediately beneath Saturn, Aristotle added three "unrolling," or counteracting, spheres for Saturn. The purpose of the three unrolling spheres was to counteract the motions of three of Saturn's four spheres, with the exception of the sphere representing the daily motion (because the daily motion was common to all planets, each was assigned a special sphere for that purpose, thus acknowledging that the daily motion was transmissible through each set of planetary spheres). As D. R. Dicks explains it:

> Thus for the four spheres of Saturn, A, B, C, D, a counteracting sphere D' is postulated, placed inside D (the sphere nearest the earth and carrying the planet on its equator) and rotating round the same poles and with the same speed as D but in the opposite direction; so that the motions of D and D' effectually cancel each other out, and any point on D will appear to move only according to the motion of C. Inside D' a second counteracting sphere C' is placed, which performs the same function for C as D' does for D; and inside C' is a third counteracting sphere B' which similarly cancels out the motion of B. The net result is that the only motion left is that of the outermost sphere of the set, representing the diurnal rotation, so that the spheres of Jupiter (the next planet down) can now carry out their own revolutions as if those of Saturn did not exist. In the same manner, Jupiter's counteracting spheres clear the way for those of Mars, and so on (the number of counteracting spheres in each case being one less than the original number of spheres in each set) down to the moon which, being the last of the planetary bodies (i.e. nearest the earth), needs, according to Aristotle, no counteracting spheres.[6]

Instead of the four spheres that Callippus required for Saturn, we see that Aristotle assigned seven. Similarly, he thought it necessary to add counteracting, or unrolling, spheres for all the planets, except the Moon, located directly above the sublunar region. Thus did Aristotle move from Callippus's system of thirty-three mathematical, or hypothetical, spheres to fifty-five physical orbs.

A momentous question was immediately posed: what caused the orbs to move around with uniform, circular motion as they carried the planets and stars? Aristotle transmitted a dual, and conflicting, legacy. In his cosmological treatise, *On the Heavens*, he appealed to an internal principle of movement when he described the celestial ether as a "simple body naturally so constituted as to move in a circle in virtue of its own nature" (2.1.284a.14–15). But in his *Physics* and *Metaphysics*, Aristotle assumed that external spiritual movers, or intelligences, were the causative agents of the rotary motions of celestial orbs. In this scheme, Aristotle assumed that each physical orb had its own immaterial mover, which, although completely immobile, was eternally able to cause its assigned orb to move effortlessly around the earth with uniform, circular motion. These "immovable," or "unmoved," movers were unique in the world because they were capable of causing motion without themselves being in motion. The potentially infinite regress of causes and effects for all motions came to a halt with the unmoved movers, which were thus the ultimate immobile sources of all motions. Although Aristotle spoke of fifty-five unmoved movers, his concept of God focused on the unmoved mover associated with the sphere of the fixed stars, the outermost circumference of the world. For Aristotle, this most remote of unmoved movers was the "prime mover," which enjoyed a special status as first among equals. Nevertheless, its role as a celestial mover differed in no way from that of all other unmoved movers, or intelligences, as they were usually called.

How did an immaterial unmoved mover cause a physical orb to move? "It produces motion by being loved" was Aristotle's response (*Metaphysics* 12.7.1072b.3–4). Precisely what he meant by this, Aristotle left unexplained. How were the motive cause and the thing moved related? Not only did his cryptic phrase tax the ingenuity of many subsequent commentators, but the intriguing thought of love as a cosmic motive force also seems to have captured the fancy of poets and jongleurs. In the last line of the *Divine Comedy*, Dante speaks of "The love that moves the sun and the other stars" (*l'amor che move il sole e l'altre stelle*),[7] and an anonymous French song proclaims "Love, love makes the world go round" (*L'amour, l'amour fait tourner le monde*).[8] If an English-language counterpart failed to appear in the Middle Ages or the Renaissance, it finally emerged in the Gilbert and Sullivan operetta *Iolanthe*, where we learn that "It's love that makes the world go round."[9] Although it is by no means certain that Aristotle is the ultimate source of these poetic sentiments, he is surely a – if not *the* – leading candidate.

Because he characterized the celestial ether as a divine and incorruptible substance and viewed terrestrial matter as the source of incessant change by means of generation and corruption, Aristotle was convinced that the unchanging celestial region exercised a dominant influence on the always changing terrestrial region. It was fitting that a nobler and more perfect thing should influence a less noble and less perfect thing. Here also was a powerful reinforcement for traditional astrological belief. The various ways in which celestial dominance was effected exercised the minds of natural philosophers until the end of the seventeenth century when the conception of the cosmos was radically altered. But as with the cause of celestial motion, Aristotle left an ambiguous legacy. Although Aristotle believed that terrestrial bodies were subject to celestial domination, he also believed that they were capable of causing effects by themselves, and were not merely passive entities dependent on celestial causes. As entities composed of matter and form, terrestrial bodies possessed natures of their own that were capable of producing effects. A heavy body fell toward the center of the earth not by virtue of any celestial power but because it possessed a nature that enabled it to do so when otherwise unimpeded. Each species of animate and inanimate being had characteristic features and properties that enabled its individual members to act in accordance with those properties.

The model for celestial activity and influence on terrestrial affairs was undoubtedly the Sun, whose influences were manifest and palpable. Its annual march around the ecliptic produced the seasons, which in turn produced various generations and corruptions. Human generation was also dependent on the Sun, as evidenced by Aristotle's widely quoted assertion that "man is begotten by man and by the sun as well."[10] With the exception of the Moon, evidence for celestial activity by the other planets was virtually nonexistent. Nevertheless, Aristotle assumed that they were also actively involved in terrestrial change. But he failed to explain how the activities of celestial bodies other than the Sun were related to the independent natures of terrestrial bodies. Once again, subsequent commentators were left to their own devices.

Many of Aristotle's major ideas and concepts about the physical world have now been described. Not only were they instrumental in shaping medieval views about the ways in which changes occurred in the terrestrial region but they also explained why these same changes were assumed not to occur in the celestial region. The ideas described here formed the core of medieval natural philosophy, and some, if not many, of those ideas would serve as springboards into new areas of thought. Aristotle's ideas provided not only a skeletal frame for medieval natural philosophy but also much of the muscle and tissue. And yet there are themes about which Aristotle provided little guidance, because either the topic was unknown to him or he had little to say about it. On other occasions, he was vague, unclear, or ambiguous, and his commentators had to work things out for themselves. At other times, his explanations were seen to be inadequate and cried out for replacement. Occasionally, his interpretations were drastically modified on the basis of experience, as with his system of concentric orbs, or on the basis of Christian theology, as with the eternity of the world. In much, if not most, of what he said, however, Aristotle's ideas were utilized as the best and most reliable guides to the comprehension of nature and its works. To medieval scholars, he was truly the Philosopher. In his commentary on Aristotle's *On the Heavens* (*De anima*), Averroes paid Aristotle the highest possible tribute, declaring that Aristotle was

> a rule and exemplar which nature devised to show the final perfection of man ... the teaching of Aristotle is the supreme truth, because his mind was the final expression of the human mind. Wherefore it has been well said that he was created and given to us by divine providence that we might know all that is to be known. Let us praise God, who set this man

apart from all others in perfection, and made him approach very near to the highest dignity humanity can attain.[11]

David Knowles, a historian of medieval philosophy, was not exaggerating when he called this "the most impressive eulogium ever given by one great philosopher to another."[12] Indeed, Averroes considered Aristotle to be virtually infallible because in over one thousand years no error had been detected in his writings.[13]

Aristotle was also greatly admired in the Latin West. Dante spoke for many when he described Aristotle as "the Master of them that know."[14] Thomas Aquinas regarded Aristotle as someone who had attained the highest possible level of human thought without benefit of the Christian faith. We might suppose that with such reverential attitudes medieval scholars would have sought to stay as close as possible to the great master. But for reasons already given, they often moved away. In chapter 6, I describe the manner in which Aristotle's medieval disciples and admirers altered and expanded his natural philosophy, even as they upheld its basic principles and remained faithful to its overall spirit. Before that, however, I shall describe the turbulent introduction of Aristotelian natural philosophy into Europe during the thirteenth century.

# 5

# The reception and impact of Aristotelian learning and the reaction of the Church and its theologians

M AJOR points of conflict existed between Church doctrine and ideas espoused in the natural books of Aristotle. The introduction of Aristotle's works into Latin Christendom in the thirteenth century was potentially problematic for the Church and its theologians. Although a clash was hardly inevitable, it was not long in coming. It seems to have hit hardest at the University of Paris, which not only had the greatest theological school of the Latin Middle Ages but also had one of the best and largest arts faculties. And yet, the conflict that developed must never be allowed to obscure the most important fact: that the translated works of Aristotle were enthusiastically welcomed and highly regarded by both arts masters and theologians. Indeed, Aristotle's philosophy was so warmly received that try as they might, the forces arrayed against it could not prevail.

## THE CONDEMNATION OF 1277

The struggle against Aristotle was concentrated in the University of Paris and its environs. In 1210, soon after Aristotle's works in natural philosophy had become available in Latin, the provincial synod of Sens decreed that the books of Aristotle on natural philosophy and all commentaries thereon were not to be read at Paris in public or secret, under penalty of excommunication. Confined to the locale of Paris, this ban was repeated in 1215, specifically for the University of Paris. On April 13, 1231, the same ban was modified and given papal sanction by Pope Gregory IX, who, in a famous bull, *Parens scientiarum* (often called, for other reasons, the Magna Carta of the University of Paris), ordered the offensive Aristotelian treatises purged of error, for which purpose he appointed a three-man commission on April 23. For reasons as yet unknown, the pope's committee failed to submit a report, and the command to expurgate the books of Aristotle was never executed. Curiously, in 1245 Pope Innocent IV extended the ban to the University of Toulouse, from whence had emanated some years earlier (1229) an invitation to masters and students to come to Toulouse where the books of Aristotle, forbidden at Paris, were openly studied. The ban on Aristotle's books on natural philosophy at Paris was in effect for approximately forty years, until 1255 at the latest. (It appears that only Aristotle's ethical and logical works were publicly taught at Paris; despite the public and private ban, the physical and philosophical works were probably read privately.) In that year, a list of texts in use for lecture courses at the University of Paris included all of Aristotle's available works. The onerous, but impractical, restrictions placed upon Parisian scholars were at an end, and they could now enjoy the same privileges as their Oxford colleagues, who were never denied the right to study and comment upon *all* the works of Aristotle during the long years of prohibition at Paris.

During the 1260s and 1270s, a second phase of the struggle developed at Paris. Inspired by Saint Bonaventure (John of Fidanza) (1221–1274), conservative theologians sought to establish limits on Aristotle's philosophy, which was the core of the new pagan and Arabic learning. The time had long passed when a simple ban on the reading of Aristotle's works could be implemented to any effect. Rather than ban works, conservative theologians sought to cope with the problem by the condemnation of ideas that they thought were dangerous and offensive. When it became apparent that their repeated warnings about the perils of secular philosophy were to no avail, the traditional theologians appealed to the bishop of Paris, Etienne Tempier, who, in 1270, intervened and condemned 13 articles that were derived either from the teachings of Aristotle or from the commentaries of Averroes on the works of Aristotle. In 1272 the masters of arts at the University of Paris instituted an oath that compelled them to avoid consideration of theological questions. If for any reason an arts master found himself unable to avoid a theological issue, he was further sworn to resolve it in favor of the faith. The intensity of the controversy was underscored by Giles of Rome's *Errors of the Philosophers*, written sometime between 1270 and 1274, in which Giles compiled a list of errors drawn from the works of the non-Christian philosophers Aristotle, Averroes, Avicenna, Algazali (al-Ghazali), al-kindi, and Moses Maimonides. When these countermoves failed to resolve the turmoil, a concerned Pope John XXI instructed the bishop of Paris, still Etienne Tempier, to investigate. Within three weeks, in March 1277, Tempier, acting on the advice of his theological advisers, issued a massive condemnation of 219 propositions.

Although the list of articles condemned by theological authorities was drawn up in haste, without apparent order and with little concern for consistency or repetition, many of them were relevant to science and natural philosophy. The condemnation of an article, however, did not of itself signify that the article was controversial in natural philosophy. The authorities may either have exaggerated its importance or simply have perceived it as potentially dangerous for public discussion. Indeed, some condemned articles may not have been expressed in writing but were perhaps only spoken, either in public disputes or in private conversations. Moreover, the inclusion of an article may even have conferred upon it an importance that it would not otherwise have had. Most of the 219 articles condemned in 1277 reflected issues that were directly associated with Aristotle's natural philosophy, and, therefore, their condemnation formed part of the reception of Aristotelian learning.

Before turning to those specific issues, however, it is essential to describe an intense interdisciplinary struggle in the thirteenth century involving the faculties of arts and theology. Essentially, the question was whether the arts faculty was entitled to equal stature with the theological faculty. The conflict expressed itself in numerous ways, but in none more basic than the overriding struggle between reason and revelation. Reason was the mode of analysis in philosophy, which was often considered co-extensive with the theoretical sciences, most of which would not themselves become independent disciplines until the seventeenth century and later. The arts masters ruled over the domain of reason and, therefore, of philosophy. But theologians held sway over revelation, and it is not difficult to understand why they held the upper hand in a society dominated by religion.

Most thirteenth-century theologians were convinced that revelation was superior to all forms of knowledge and therefore subscribed to the traditional doctrine of secular learning as the handmaiden to theology. One of the most important of these thirteenth-century theologians, Saint Bonaventure, devoted an entire treatise to the proposition that the secular subjects taught in the arts faculty at the University of Paris were subordinate to the discipline called theology taught in the faculty of theology. In his treatise *Retracing the Arts to Theology* (*De reductione artium ad theologiam*), Bonaventure sought to show that theology is the queen of the sciences, because, in the final analysis, all learning and knowledge depend upon divine illumination from Sacred Scripture, the study of which is the exclusive domain of theologians. In Bonaventure's world, and that of many theologians, faith and reason were harmoniously unified, with the former ultimately guiding and informing the latter.

The teachers in the arts faculties of Paris and elsewhere had a radically different view of the relationship of their discipline to theology. In the broadest sense, they taught philosophy, which, although it included the seven liberal arts as introductory subjects, primarily consisted of metaphysics, natural philosophy, and moral philosophy. Because philosophy as a whole was based overwhelmingly on the writings of Aristotle, most arts professors thought of themselves as followers of Aristotle and regarded him as the embodiment of reasoned analysis. Indeed, their livelihoods were based upon the explication of Aristotle's ideas and thoughts. As a mark of their respect for him, medieval scholastic authors usually referred to Aristotle by the honorific title "the Philosopher" (*philosophus*). They regarded themselves as the guardians of reason and took pride in their role as philosophers. If left to themselves arts masters would probably have applied reason to all branches of knowledge, including theology. Indeed, many of them would have followed reason to its conclusions, even where this conflicted with revelation, although, in the end, they would have yielded to revelation on the basis of faith. Nevertheless, they regarded philosophy as the proper tool for understanding the world. For them, it was worthy of independence from theology, and they fought for its autonomy (for more on this, see chapter 8). Although theologians were themselves very interested in philosophy (and natural philosophy), and many believed it was a discipline distinct from theology, most assigned it subordinate status. During the thirteenth century, the first century of the institutionalization of Aristotelian natural philosophy in Western Europe, tensions between these two university disciplines and their separate faculties were almost inevitable.

The dispute is apparent in at least three major controversies, which involved (1) the eternity of the world, (2) the so-called doctrine of the double truth, and (3) God's absolute power. Interdisciplinary friction between theologians and natural philosophers was compounded by intradisciplinary rivalries among the theologians themselves. Neoconservative Augustinians were pitted against the Dominican followers of Saint Thomas Aquinas, with the former concerned about the overreliance of the Dominicans on Aristotelian philosophy and the latter determined to pursue a harmonization of reason and revelation. The condemned articles themselves, however, serve as good illustrations of the controversies that loomed large in the late thirteenth century.

The following three articles convey the flavor of the hostility between arts masters and theologians:

152. That theological discussions are based on fables.
153. That nothing is known better because of knowing theology.
154. That the only wise men of the world are philosophers.

If arts masters held such opinions, and some apparently did, we can appreciate the sense of outrage and the animosity that the theologians felt toward them. From the 1220s, and probably even before, Church authorities worried that philosophy was rapidly penetrating, and perhaps dominating, theology. Pope Gregory IX sought to preserve the traditional relationship between theology and philosophy, with the latter serving as handmaiden to the former. Indeed, Gregory reflected a wide concern, going back to the Church fathers, that efforts to buttress faith with natural reason were potentially dangerous because they implied that somehow faith could not stand alone. In 1228, Gregory ordered the theological masters at Paris to exclude natural philosophy from their theology.

Gregory's ban did not prevail. Not only was philosophy gradually recognized as an autonomous discipline, with Aristotle its fundamental authority just as the Church fathers were authorities in theology, but strictures against the use of natural philosophy in theology also faded, although they were revived from time to time, always in vain. Perhaps more than anyone, Thomas Aquinas sought to define the relationship between theology and philosophy. He did so by making each an independent science. The fundamental principles of theology are the articles of faith, whereas the basic principles of philosophy are founded in natural reason. Articles of faith, therefore, cannot be demonstrated by reason. If theology and philosophy are independent sciences, does it follow that those who engage in natural philosophy should not theologize and those who do theology should not philosophize? With respect to theology, Thomas believed that a theologian should use logic, natural philosophy, and metaphysics to the extent necessary, although he would not have approved of theologizing in philosophy. By establishing theology as an independent science, Thomas implicitly conceded autonomy to philosophy (and, therefore, also to natural philosophy) as a science, although he still regarded it as subordinate to theology. In the conflict that began between philosophy and theology in the thirteenth century, theology held the upper hand. Until the seventeenth century, undemonstrated, revealed truths of faith had ultimate priority over demonstrated truths of reason.

### The eternity of the world

During the 1260s some arts masters, or philosophers, were already exercising autonomy in their discipline by reasoning solely in terms of natural principles. But it was difficult to be indifferent to the theological impact of their conclusions, as can be seen in the first of the three issues mentioned earlier, namely, the eternity of the world. This issue was to relations between science and religion in the Middle Ages what the Copernican heliocentric theory was in the sixteenth and seventeenth centuries and the Darwinian theory of evolution in the nineteenth and twentieth centuries.

Following from his arguments at the end of the first book of *On the Heavens*, Aristotle concludes, at the very beginning of the second book, that "the world as a whole was not generated and cannot be destroyed, as some allege, but is unique and eternal, having no beginning or end of its whole life."[1] Because Aristotle based his natural philosophy on the firm conviction that the world is eternal, it posed a major threat to the creation account in Genesis. As evidence that the eternity of the world was regarded as potentially dangerous, 27 of the 219 articles condemned in 1277 (more than ten percent) were devoted to its denunciation. The eternity of the world was thus manifested in numerous ways. For example, article 9 condemned the proposition that "there was no first man, nor will there be a last; on the contrary there always was and always will be the generation of man from man"; article 98 condemned the proposition that "the world is eternal because that which has a nature by [means of] which it could exist through the whole future [surely] has a nature by [means of] which it could have existed through the whole past"; and article 107's claim that the elements are eternal but that "they have been made [or created] anew in the relationship which they now have" was also condemned.

Because the theological authorities condemned the eternity of the world in twenty-seven different versions, we might expect to find that belief in the world's eternity was widespread. In fact, no one has yet been identified who held this heretical opinion without qualification. Why, then, did the authorities condemn twenty-seven articles to prevent the dissemination of a proposition that no one seems to have explicitly advocated? Although it is possible that some or all of these propositions were held in private and that this was common knowledge, a more likely reply lies in the nature of the responses to claims about the eternity of the world, as is evident in the reactions of the two best-known arts masters of the thirteenth century, Boethius of Dacia (d. after 1283) and Siger of Brabant (d. ca. 1284), both of whom fled France for Italy after the promulgation of the Condemnation of 1277.

Boethius and Siger each wrote a treatise on the eternity of the world, and Boethius also considered it in his *Questions on the Physics* (*Quaestiones super libros Physicorum*). In his treatise *On the Eternity of the World* (*De aeternitate mundi*), Boethius argues that no philosopher could demonstrate that a first motion had ever come into being, and hence a beginning of the world is not determinable. The eternity of the world, however, is no more demonstrable than is its creation. Although no acceptable proof could be proposed for either position, Boethius insists that there is no contradiction between the Christian faith and philosophy. The faith must prevail. He concludes that

> the world is not eternal, but was created anew, although . . . this cannot be demonstrated by arguments, just as may be said about other things that pertain to the faith. For if they could be demonstrated, they would not belong to the faith, but to science. . . . There are many things in the faith which cannot be demonstrated by reason, as [for example] that a dead person comes to life again numerically the same as he is now, and that a generable thing returns without generation. And who does not believe these things is a heretic; [and] whoever seeks to know these things by reason is a fool.[2]

In his *Questions on the Physics*, however, written at approximately the same time, Boethius argues that prime matter is eternal and therefore has to be coeternal with God. Indeed, God has to be viewed as the creator of prime matter. For Boethius, this conclusion followed logically from the application of reason to the operations of the world. In this context, God is still perceived as creator of both matter and world, but the "created" matter is nevertheless eternal.

Siger argued similarly. The world and its species cannot have been created, because no species of being could have been actualized from a previous state of potentiality, and therefore every species must have been in existence previously. Although reason led him to this conclusion, which seemed to proclaim the eternity of the world, Siger sought to protect himself against possible charges of heresy by insisting that "we say these things as the opinion of the Philosopher [i.e., Aristotle], although not asserting them as true."[3] Where the pronouncements of faith conflicted with Aristotle's conclusions, the faith must prevail.

The attitude of Boethius and Siger was probably typical of that of other – perhaps of many – of the arts masters of the late thirteenth century and was exhibited, in the fourteenth century, by John of Jandun, a famous and controversial arts master. When Church doctrine conflicted directly with the conclusions of Aristotelian natural philosophy – as it did on the question of the eternity of the world – the arts masters yielded to theology and faith. Indeed, as we saw, the arts masters of Paris were compelled to do so by oath as far back as 1272, a requirement that remained in effect until the fifteenth century.

Even the theologians were in serious disagreement. One of the most important among them, Thomas Aquinas, broke with his conservative colleagues and adopted an approach similar to that of Boethius of Dacia. Like Boethius, Thomas denied that any proper demonstration could be formulated in favor of creation or eternity. Therefore, one has to concede that the eternity of the world is a possibility (for Thomas's arguments, see chapter 6). To the bishop of Paris and like-minded traditional theologians, the arguments proposed by Boethius, Siger, and Thomas must have been suspect. They appeared to confer respectability on belief in the eternity of the world, even as they seemed to undermine confidence in its creation. Nevertheless, on the basis of faith, all three proclaimed their belief in the creation of the world as described in Genesis. As Thomas put it: "That the world had a beginning . . . is an object of faith, but not of demonstration or science."[4]

### The doctrine of the double truth

When arts masters yielded to the faith, however, they appeared to do so in a manner that left the theologians uneasy and suspicious. They implied, and often explicitly stated, that the truths of natural philosophy, based on the application of natural reason to a priori principles and sense experience, could not be reconciled with the truths of faith. Under such circumstances, the faith had to be upheld. But it was upheld ambiguously because the arts masters usually left the reasoned conclusions of natural philosophy intact, even as they proclaimed the corresponding truths of the faith. If the eternity of the world, for example, was considered an appropriate conclusion in natural philosophy, it was nevertheless contrary to the faith and therefore had to be rejected. Under such circumstances, it was obvious that the arguments in favor of the eternity of the world had not been rejected because they were flawed, but only because they were contrary to the faith. It left the impression that there were two truths, one for natural philosophy, and one for the faith. Because arts masters usually refrained from reconciling Aristotle's principles and conclusions – in which they presumably believed – with the truths of faith, they seemed to be subtly advancing the cause of Aristotle. At the very least, they appear to have conveyed an impression to the theologians that they subscribed to a doctrine of double truth, as is evidenced in the Condemnation of 1277. In his prologue to the condemnation, the bishop of Paris briefly mentions a doctrine of double truth when he denounces those who say that "things are true according to philosophy, but not according to the Catholic faith; as if there could be two contrary truths."[5] As an exemplification of his meaning, the bishop could point to article 90, which condemned those who believed that "a natural philosopher ought to deny absolutely the newness [that is, the creation] of the world because he depends on natural causes and natural reasons. The faithful, however, can deny the eternity of the world because they depend upon supernatural causes."

Although some arts masters appear to have come close to implicit acceptance of a double truth, no arts master has yet been discovered who believed literally in a doctrine of the double truth. But, on the basis of what has already been said, we can understand why many theologians may have thought that Boethius of Dacia, Siger of Brabant, and others – even including one of their own, Thomas Aquinas – actually believed in the eternity of the world, even as they proclaimed their fidelity to the Christian dogma of creation. All this is made vividly apparent in Armand Maurer's description of Boethius of Dacia's approach to the eternity of the world:

> For there to be two contrary truths, the Christian truth that the world is not eternal would have to be opposed to a philosophical truth that the world is eternal. But we look in vain in Boetius' treatise for the statement that the eternity of the world is philosophically true. We are told simply that it follows from the principles of natural philosophy. In one place Boetius asserts that it follows from the "truths of natural causes"; but the

conclusion itself is not explicitly said to be true. Boetius comes so close to affirming a two-fold truth at this point, and yet avoids it so adroitly, that we can only conclude that he did so deliberately. Like Siger of Brabant, he appears to be very careful not to bring faith and philosophy into open contradiction in the realm of truth. And yet he comes so close that we can see why he was condemned by the Bishop of Paris.[6]

## Limitations on God's absolute power

Of the three major issues described earlier, the third, the challenge to God's absolute power, may have been regarded as the most potentially subversive of theological traditions. Scattered through the works of Aristotle were propositions and conclusions demonstrating the natural impossibility of certain phenomena. For example, Aristotle had shown that it was impossible for a vacuum to occur naturally inside or outside the world and had also demonstrated the impossibility that other worlds might exist naturally beyond ours. Theologians came to view these Aristotelian claims of natural impossibility as restrictions on God's absolute power to do as he pleased. Why should God not be able to produce a vacuum inside or outside the world if he chose to do so? Why could he not create other worlds if he chose to do so? Article 147 reveals the attitude of the bishop of Paris and his colleagues when it denounced as erroneous the opinion that not even God could do what was naturally impossible. The following articles in the Condemnation of 1277 were among those thought to place limitations on God's absolute power to do whatever he pleases:[7]

21. That nothing happens by chance, but all things occur from necessity and that all future things that will be will be of necessity, and those that will not be it is impossible for them to be. . . .
34. That the first cause [that is, God] could not make several worlds.
35. That without a proper agent, as a father and a man, a man could not be made by God [alone].
48. That God cannot be the cause of a new act [or thing], nor can He produce something anew.
49. That God could not move the heavens [or world] with a rectilinear motion; and the reason is that a vacuum would remain.
139. That an accident existing without a subject is not an accident, except equivocally; [and] that it is impossible that a quantity or dimension exist by itself because that would make it a substance.
140. That to make an accident exist without a subject is an impossible argument that implies a contradiction.
141. That God cannot make an accident exist without a subject, nor make several dimensions exist simultaneously [in the same place].

Many more articles limiting God's power could be cited. All were condemned because the theological authorities wanted everyone to concede that God could do anything whatever short of a logical contradiction. By condemning the opinion that God could not create other worlds, article 34 made it mandatory to concede that he could create as many as he pleased. Although no one was required to believe that God had created other worlds, and no one is known to have so believed, the effect of article 34 on natural philosophy was to encourage speculation about the conditions and circumstances that would obtain if God had indeed created other worlds. Article 49 denied to God the capability of moving the outermost heaven, and therefore the world itself, with a rectilinear motion, because such a motion would leave behind a vacuum after the departure of the world from its present position. Following the condemnation of article 49 in 1277, scholastic natural philosophers routinely conceded that if it pleased God to do so, he could indeed move the world rectilinearly.

In articles 139, 140, and 141, the authorities condemned the seemingly self-evident Aristotelian principle that an accident could not exist without a subject, or substance, in which to inhere and that God could not create an accident, or quality, that did not inhere in a subject or substance. They further condemned (in articles 139 and 141) the Aristotelian axiom that no quantity or dimension could exist independently of a material body and also denounced the equally fundamental Aristotelian principle that two or more dimensions could not exist simultaneously in the same place. Not only did articles 139, 140, and 141 qualify as placing limitations on God's power, but they denied to God the power to effect the theological dogma of the Eucharist, whereby God miraculously produces the actual body and blood of Christ in the bread and wine used in the mass. Although the bread has become the substance of the body of Christ, the bread itself remains with its usual visible, accidental properties. Those accidents, however, no longer inhere in the substance of the bread. And because their existence in Christ is denied, one must conclude that the accidents of the bread do not inhere in any substance, a conclusion that is contrary to Aristotle's doctrine that all accidents must inhere in a substance. In the fourteenth century, Walter Burley argued that just as (in the Eucharist) God makes a quantity devoid of corporeal substance, so also could he make a quantity devoid of inhering qualities, by which Burley meant an extended vacuum through which light and heavy bodies could move. Thus did Burley link the condemned articles on the supernatural separation of accidents from their subjects, or attributes from their substances, with a much-discussed medieval problem of motion in a separate, empty space.

Article 48 was also condemned because it placed restrictions on God's absolute power. The concept of a lawful, regular, and even deterministic world held appeal for astrologers and was also basic to Aristotle's natural philosophy. The idea that God would intervene miraculously in the natural order and do something new or make something new was alien to Aristotle's physics but was an important feature of the Christian religion. Medieval followers of Aristotle's natural philosophy were now warned that they could not deny the occurrence of such actions.

Other articles condemned in 1277 also played a role in the physical and cosmological discussions of the fourteenth century. Many were relevant to one of the three major themes of the second half of the thirteenth century that we have already discussed: the eternity of the world, the doctrine of the double truth, and God's absolute power.

### TWO SENSES OF THE HYPOTHETICAL IN MEDIEVAL NATURAL PHILOSOPHY

Although natural philosophy was primarily concerned with the real physical and intelligible world, hypothetical arguments played a significant role. Two kinds may be conveniently distinguished: one that operated in defense of Aristotle's principles of natural philosophy, and another that operated against those principles and was influenced by the Condemnation of 1277.

Natural philosophers, who regularly considered Aristotle's explicitly heretical principles and who also confronted religious dogmas that were often inconvenient for arguments in natural philosophy, were fully cognizant of the need to avoid the appearance of heresy or face dire consequences. Principles and ideas often had to be recast into hypothetical language or proper qualifications had to be added. This was usually done under the guise of "speaking naturally" (*loquendo naturaliter*). One could proclaim the eternity of the world when "speaking naturally" or could assume that every accident and quality, without exception, inhered in a subject or substance. When "speaking naturally," authors or lecturers were understood to be focused only on natural philosophy, without regard for theological implications or complications. They had to resort to the kinds of subtleties and distinctions that were invoked by Siger of Brabant and Boethius of Dacia, as described earlier in this chapter. This was a standard procedure. In the fourteenth century, for example, Albert of Saxony, an influential scholastic arts master and natural philosopher, assumed the truth of the eternity of the world and also the existence of a fixed quantity of matter in the world.[8] From these assumptions, he concluded that over an infinite time, this limited quantity of matter would have to furnish bodies for an infinite number of souls. In the course of an eternal period of time, the same matter might serve as the human body of a number of different souls. On the day of resurrection, however, when every soul receives its own material body, a finite quantity of matter would have to receive an infinite number of souls. This was a heretical state of affairs, because one body – indeed every body –

would have to receive more than one soul. Albert's response to this dilemma was typical for natural philosophers who had to contend with theological restrictions. He explains that "the natural philosopher is not much concerned with this argument because when he assumes the eternity of the world, he denies the resurrection of the dead."[9] In this instance, Albert simply dropped the inconvenient theological consequences from his discussion but retained the eternity of the world for the sake of the argument. By such appeals and devices, medieval natural philosophers could assume the truth of almost any condemned proposition, provided that they did not proclaim it to be categorically or philosophically true. In this way, Aristotle's controversial ideas and principles were maintained and analyzed throughout the Middle Ages, with few major repercussions.

In the example just cited, Albert of Saxony was eager to drop the theological aspects of his discussion. Occasionally, however, situations arose in which a natural philosopher believed that the theological implications of a problem were sufficiently important to include them as part of a proper resolution of the issue. For example, in his analysis of the vacuum, which he believed touched faith and theology, John Buridan insisted that it was essential to present the theological arguments so that he might resolve them in favor of the faith, as required by the oath sworn by masters of arts since 1272. Buridan felt compelled to mention this because, as he put it, "some of my lords and masters in theology have reproached me on this, [saying] that sometimes in my physical questions I intermix some theological matters which do not pertain to the artists [that is, masters of arts]."[10] Buridan knew, however, that he had the right to introduce theological material, provided that he resolved all outstanding problems in favor of the faith.

Of the two types of hypothetical arguments mentioned earlier, the first shows how medieval natural philosophers managed to cope successfully with Aristotle's ideas that were contrary to the faith or were, in some sense, subversive of it; the second, which derives from the impact of the concept of God's absolute power as expressed in the Condemnation of 1277, reveals a method by which natural philosophers transcended the bounds of Aristotle's confining principles and broke free to consider possibilities they might not otherwise have contemplated.

By emphasizing God's absolute power to do anything short of a logical contradiction, the articles condemned in 1277 had a curious, and probably unintended, effect: they encouraged speculation about natural impossibilities in the Aristotelian world system, which were often treated as hypothetical possibilities. The supernaturally generated alternatives, which medieval natural philosophers considered in the wake of the condemnation, accustomed them to consider possibilities that were beyond the scope of Aristotle's natural philosophy, and often in direct conflict with it. The contemplation of hypothetical possibilities that were naturally impossible in the Aristotelian world view was so widespread that speculation about them became an integral feature of late medieval thought.

Of all the themes of natural philosophy influenced by the Condemnation of 1277, with its overriding theme of God's absolute power, none was affected more than the concept of vacuum, the very idea of which Aristotle thought absurd and impossible. None could deny that the possible existence of vacua had powerful implications for theology. As Gregory of Rimini (d. 1358) rightly declared, every Catholic had to concede the possibility that God could create one. Indeed, creation itself raised a fundamental question about the vacuum: did God require an empty space in which to create the world? An argument by Averroes presented Christians with a dilemma: either concede that the world is eternal, or accept the existence of an eternal precreation void space that God required in order to have a place in which to create the world. The need for a pre-creation void space was condemned in 1277 by article 201, which proclaimed that "before the generation of the world, there was a place without a thing located in it, which is a vacuum."[11]

The concept of vacuum was also inherent in articles 34 and 49, which dealt, respectively, with the possibility of other worlds and with the rectilinear motion of our own world, both natural impossibilities in Aristotle's cosmos. The two articles generated serious discussion about the possible existence of void space beyond our world. Although no articles of the Condemnation of 1277 were directed at the possibility of vacua within our cosmos, it was obvious that if God could create a vacuum beyond the world, he surely could do so within the world. And so it was that after 1277 God was frequently imagined to annihilate all or part of the matter that existed in the material plenum of our world. Within this now empty space, many different situations were imagined for further discussion. Would the surrounding celestial spheres immediately collapse inward as nature sought to prevent formation of a vacuum? Would the empty interval be a vacuum or space? If all bodies within the concave surface of the last sphere were annihilated, so that no matter remained, would it still be meaningful to describe that concave surface as a place, even though it would be a place without body? Could a stone placed within such a vacuum be capable of rectilinear motion? Would it be possible to measure distances within such a vacuum? If people were somehow located in this vacuum, would they be able to see and hear each other? And, finally, if, as was commonly believed, celestial bodies governed the behavior of bodies in the sublunar region, what would happen to a spherically shaped piece of earth that was located in the air enclosed within a house if God destroyed everything outside that house, including the celestial orbs and all bodies? During the late Middle Ages, analysis of these and similar imaginary examples was usually made in terms of Aristotelian principles, even though the conditions imagined were "contrary to fact" and impossible within Aristotelian natural philosophy.

Thus did the concept of God's absolute power become a convenient vehicle for the introduction of subtle and imaginative questions, which often generated novel answers. Although these speculative responses did not lead to the overthrow of the Aristotelian world view, they did, as we shall see, challenge some of its fundamental principles and assumptions. They made many aware that things might be quite otherwise than were dreamt of in Aristotle's philosophy.

As far as can be determined, the Condemnation of 1277 was never formally annulled either by the bishop of Paris or by any pope. In 1325, however, two years after Thomas Aquinas was made a saint, all of the condemned articles that had been held by Saint Thomas were formally nullified, leaving the remaining articles still in effect. Through the rest of the Middle Ages, and even into the seventeenth century, one or another of the condemned "articles of Paris" was cited by theologians and natural philosophers. But was the Condemnation of 1277 a significant influence in medieval natural philosophy? Did its articles give birth to modern science, as Pierre Duhem, the great pioneer investigator into medieval science, claimed? Or were they irrelevant to the advent of modern science, as Alexandre Koyré, an eminent historian of the Scientific Revolution, would have it?

Neither of these two assessments is tenable. Although the truth may lie somewhere between these extremes, it is elusive and may be ultimately indeterminable. We can be certain only that the condemnation expanded the horizons of Aristotelian natural philosophers and made medieval natural philosophy more interesting than it would otherwise have been, and even produced a few significant surprises. The "natural impossibilities" that were explored as a consequence of the condemnation represented additions to natural philosophy, but they did not alter the main body of that discipline. They did not revolutionize Aristotelian natural philosophy nor cause it to be abandoned.

## THE THEOLOGIAN-NATURAL PHILOSOPHERS

Although during the thirteenth century some theologians feared the impact of Aristotle's natural philosophy and sought first to ban his works, then to expurgate them, and finally to condemn certain of his ideas that restricted God's absolute power, it would be a serious error to suppose that theologians opposed Aristotelian natural philosophy. Theologians differed widely among themselves on many issues and were by no means in agreement in their attitude toward Aristotle. But even the most conservative of them, like Saint Bonaventure, recognized the enormous utility of Aristotle's natural philosophy. Far from opposing it, most theologians were among its staunchest supporters. So important was natural philosophy to theologians that a high level of competence in that subject, usually in the form of a master of arts degree, was presupposed for those who entered upon the formal study of theology.

With theology and natural philosophy related so intimately during the Middle Ages, and with arts masters forbidden to apply their knowledge to theology, it remained for the theologians to interrelate these two disciplines, that is, to apply science to theology and theology to science. Because they were usually thoroughly trained in both disciplines, medieval theologians were able to interrelate natural philosophy and theology with relative ease and confidence, whether this involved the application of science and natural philosophy to scriptural exegesis, the application of God's absolute power to hypothetical possibilities in the natural world, or the frequent invocation of scriptural texts to support or oppose scientific ideas and theories. Theologians had remarkable intellectual freedom and rarely permitted theology to hinder their inquiries into the physical world. If there was any temptation to produce a "Christian science," they successfully resisted it. Biblical texts were not employed to "demonstrate" scientific truths by blind appeal to divine authority. When Nicole Oresme inserted some fifty citations to twenty-three different books of the Bible in his *On the Configurations of Qualities and Motions*, a major scientific treatise of the Middle Ages, he did so only as examples, or for additional support, but in no sense to demonstrate an argument.

If the theologians had chosen to oppose Aristotelian learning as dangerous to the faith, it could not have become the focus of studies at the university. But they had no reason to oppose it. Western Christianity had a long-standing tradition of using pagan thought for its own benefit. As supporters of that tradition, medieval theologians treated the new Greco-Arabic learning in the same manner – as a welcome addition that would enhance their understanding of Scripture. The positive attitude of medieval theologians toward natural philosophy, and their belief that it was also a useful tool for the elucidation of theology, must be viewed as the product of an attitude that was developed and nurtured during the first four or five centuries of Christianity.

The Western approach toward science and natural philosophy eventually outgrew its handmaiden attitude. Even before this occurred, however, we may plausibly assume that many, perhaps most, scholastic authors in the arts faculty enjoyed the study of science and Aristotelian natural philosophy for its own sake. Their primary concern was for natural philosophy, because they had not studied theology and were forbidden by statute from seriously discussing it. Excluded from the domain of theology, arts masters had a professional interest in natural philosophy. It is equally plausible to assume that many theologians, who wrote commentaries on Aristotle's natural books and used natural philosophy regularly in theology, may have enjoyed natural philosophy for its own sake, and those theologians who actually engaged in scientific activity probably derived some degree of pleasure from their efforts as well, whatever the implications for theology. We may reasonably surmise that because of the positive attitudes of natural philosophers and theologians toward science and natural philosophy these disciplines were highly respected. They had long since ceased to require justification for their existence and study.

The unusual development that produced a whole class of theologian-natural philosophers serves as a key to understanding the fate of science and natural philosophy in Western Europe during the Middle Ages. The amazing lack of strife between theology and science is attributable to the emergence of theologian-natural philosophers who were trained in both natural philosophy and theology and were therefore able to interrelate these disciplines with relative ease. They were able to do so because, in large measure, Christianity had early on adjusted to Greek secular thought. Occasional reactions against natural philosophy – as in the early thirteenth century, when Aristotle's works were banned for some years at Paris, and in the later thirteenth century, when the bishop of Paris issued the Condemnation of 1277 – thus become relatively minor aberrations when viewed against the grand sweep of the history of Western Christianity.

Aristotelianism was much broader than the works of Aristotle and the Latin (and much earlier, Greek and Arabic) commentaries that they generated. Aristotle's natural philosophy was imported into theology, especially into theological commentaries on the *Sentences* of Peter Lombard. It was also integrated into medicine, where the medical works of Avicenna were already heavily impregnated with Aristotelian ideas and then further embellished by physician-commentators who were thoroughly acquainted with Aristotle's natural philosophy. Music theorists also found it occasionally convenient to introduce concepts from natural philosophy to elucidate musical themes and ideas. Largely because Aristotle's works formed the basis of the medieval university curriculum, Aristotelianism emerged as the primary, and virtually unchallenged, intellectual system of Western Europe. It not only provided the mechanisms of explanation for natural phenomena but also served as a gigantic filter through which the world was viewed. In the next chapter, we shall describe the manner in which medieval natural philosophers responded to particular aspects of their legacy of Aristotle's natural books and the departures that they made from it.

# 6

# What the Middle Ages did with its Aristotelian legacy

MANY of the Aristotelian principles and concepts described in chapter 4 were retained in the Middle Ages. These concepts – element, compound, matter, form, the doctrine of contraries, the four types of change, celestial incorruptibility, and others – were too fundamental to be abandoned or even altered significantly. Numerous other of Aristotle's arguments and ideas, however, were changed considerably and sometimes wholly replaced. The most significant transformations in Aristotle's natural philosophy during the late Middle Ages, from the standpoint of the history of science, occurred in his treatment of motion. Here the departures were truly dramatic. Aristotle's explanations of natural and violent motion were largely abandoned, especially those on violent motion.

One striking departure occurred in Aristotle's basic formulation as described in chapter 4. There we saw that for Aristotle $V \propto F/R$, where $V$ is velocity, $F$ is motive force, and $R$ is the total resistance offered to the applied force, a quantity that, presumably, includes the resisting object or body plus the resistance of the external medium in which the motion occurs. To double $V$, $R$ could be halved and $F$ held constant; or, $F$ doubled and $R$ held constant. To halve $V$, $F$ could be halved and $R$ held constant; or $R$ doubled and $F$ held constant. Aristotle realized that in halving a velocity $F$ might be reduced to the point where it was smaller than $R$ and therefore unable to move it. Under these circumstances, he insisted that the rules of motion would no longer apply, and movement would cease immediately.

His critics realized that, in order to preserve his law of motion, Aristotle had accepted a discontinuity and had broken the connection between the physical process and the continuous mathematical function. They argued that in halving any velocity, the force, $F$, could be successively halved, or $R$ could be successively doubled until $F$ was equal to, or less than, $R$, at which point it was self-evident that motion would cease. And yet, mathematically, the Aristotelian function $V \propto F/R$ indicates a positive velocity, because $F/R$ is representable by a fraction, however small. Aristotle's position seemed to commit its supporters either to the physically absurd position that any force, however small, could move any resistance, however large, or to a mathematical representation that continued to register velocities after they were physically impossible to produce.

To avoid this dilemma, Thomas Bradwardine (ca. 1290–1349), in 1328, proposed a new mathematical relationship based on geometric proportionality, which came to be called "ratio of ratios." In this new function, or "law of motion," to halve a velocity produced by a ratio of force to resistance, it was necessary to take the square root of the ratio $F/R$, or $(F/R)^{1/2}$. On this approach, if $F$ is initially greater than $R$, and motion occurred, $F$ could not become equal to or less than $R$, because the reduction of $V$ by half is no longer achieved by halving $F$ or doubling $R$. To reduce a velocity by one third, one had to take the cube root of $F/R$, or $(F/R)^{1/3}$, and so on. The doubling or tripling of velocities was produced by squaring or cubing $F/R$, that is, by generating a ratio equal to $(F/R)^2$ or $(F/R)^3$, respectively. No longer was this accomplished by doubling or tripling $F$ alone, or by taking the one-half or one-third part, respectively, of $R$ alone.

In the ratios described in the preceding paragraph, the $R$, or resistance, included both the weight of the body being set in motion and the resistant medium through which it moved. A major criticism developed over Aristotle's concept of a resistant medium. In a plenum, what role did the matter that filled the sublunar world play? Was it essential for motion, or was motion possible without it? Here was a formidable problem, one that had already engaged the attention of earlier commentators.

## THE TERRESTRIAL REGION

Long before Aristotelian physical science reached the Latin West in the twelfth and thirteenth centuries, Greek and Arabic commentators had produced a body of literature in which local motion was intensively discussed. Occasionally, significant criticisms were raised and certain of Aristotle's opinions called into question, as when John Philoponus, a Greek commentator of the sixth century A.D., disputed the role that Aristotle had assigned to the external medium. Not only did Philoponus deny the necessity for a resistant medium in local motion, but he also rejected the external medium, especially air, as the agent or cause of violent motion and suggested instead an incorporeal, impressed force. Arab commentators, familiar with the works of some of the Greek commentators, frequently elaborated on and added to these ideas, some of which reached medieval Europe in Latin translation. Averroes, for example, transmitted a brief anti-Aristotelian critique by Avempace (the Latinized form of Ibn Bajja), an Arab who lived in Spain and died in 1138, who may have been influenced by Philoponus.

In his commentary on Aristotle's *Physics*, Averroes reports that Avempace denied Aristotle's claim that the time of fall of a body is directly proportional to the density, and therefore the resistance, of the external medium through which it falls. Aristotle's claim would be true, Avempace argued, only if the time required to move from one point to another was due solely to the resistive capacity of the intervening medium. On this crucial point, Aristotle himself furnished Avempace with a powerful counterargument. Despite his observation that planets and stars, like terrestrial bodies, do not move instantaneously from one point to another, Aristotle had also insisted that heavenly bodies move effortlessly through a resistanceless celestial ether. Because it was obvious from observation that planets moved around the heavens with different periodic motions, it followed that differing finite planetary speeds could occur without the active resistance of a medium. Avempace concluded not only that a resistant medium was unessential for the occurrence of motion but that the sole function of a material medium was to retard motion. The ordinary observable motion of a body in a medium is retarded motion caused by the resistance of the medium. Avempace inferred that in the absence of a resistant medium, a body would necessarily fall with a faster, natural speed. This hypothetical unobstructed speed would be reduced in proportion to the retardation caused by the resistance of the medium.

The vagueness of Avempace's account – or, perhaps more accurately, the vagueness of Averroes's report of his account – made actual determination of observable motion impossible. Avempace suggested no means by which motion in a resistanceless medium, or vacuum, might be measured. Should it be measured by the body's weight, by its dimensions, by an indwelling power, or in some other way? Precisely how would the total resistance of the medium retard natural motion and produce the final velocity of an observable motion? How should such resistance be measured? Not until the sixteenth century, when Giovanni Battista Benedetti and Galileo adopted similar anti-Aristotelian positions, was a genuine effort made to provide an objective measure for the resistance of a medium.

Soon after the works of Averroes became available in Latin translation, Avempace's critique became widely known, giving rise to further controversy. Thomas Aquinas (1225–1274) was one of the first to consider it. Though he did not mention Avempace by name, his succinct argument against Aristotle and Averroes left no doubt of his pro-Avempace viewpoint. As empirical evidence that motion in a resistanceless medium would be finite, Thomas repeated Avempace's illustration of motion through the celestial ether, an illustration that soon became commonplace. But reason also tells us that motion in a vacuum would be finite and successive, because void space, no less than space filled with matter, is an extended, dimensional magnitude. To move from one distinct point to another, a body must traverse the intervening empty or full space. To do this, it must first pass through the parts of space that are nearer the starting point before it reaches those parts that are farther from it. If a body could move through a vacuum in this way, some scholastic authors suggested that by virtue of its extension, the vacuum itself functioned as a resistance, because the purpose of a resistance was to cause a body to move from one point to another in a finite time. Attempts to conceive of void space as a resistance simply because voids were three-dimensional and might be traversed sequentially failed. Natural philosophers were apparently reluctant to believe that pure emptiness could function as a physical resistance in the same manner as air, or water. In purely spatial and temporal terms, however, it appeared that motion in a vacuum could be finite and also successive, rather than infinite and instantaneous, as Aristotle had claimed. Kinematically, at least, motion in a vacuum was intelligible and seemed feasible.

### The causes of motion

In the Aristotelian physical world, however, bodies did not just move in space and time. Their motions had to be explained causally. Could motion in void spaces be accounted for by the usual dynamic principles associated with the ordinary motions of physical bodies? If a real body were placed in a real vacuum, assuming that one existed, would it rise or fall with a natural motion? If it were hurled away from, or out of, its natural place, could it move with a continuous, violent motion? Although Aristotle had repudiated the possibility of motion in a vacuum and could furnish no guidelines to those who raised such questions, the medieval response was nevertheless formulated with Aristotelian physical principles firmly in mind. Natural philosophers therefore assumed that whatever is moved is moved by a distinct, separate, and identifiable entity, and that every motion involves the action of a force operating against a resistance. If the vacuum itself played no role as either motive force or resistance, what might perform these functions? What could count as a motive force, and what might be identifiable as a resistance?

*Internal resistance and natural motion in a vacuum.* The solution with regard to natural motion followed upon the introduction, sometime in the late thirteenth or early fourteenth century, of a new concept, *internal resistance*. This was made possible by a new interpretation of Aristotle's notion of a mixed, or compound, body. Aristotle had distinguished pure elemental bodies (earth, air, water, and fire), which were mere hypothetical entities not actually observed in nature, from compound, or mixed bodies, which were mixtures, in varying proportions, of all four elements and were the bodies actually observed in nature. Aristotle held that in every mixed body one of the elements would dominate and determine its natural motion, that is, determine whether the body would naturally rise or fall. Although this interpretation remained acceptable to many during the Middle Ages, some came to believe not only that a mixed body could be composed of two, three, or four elements but also that a predominant element did not determine its natural motion. Rather, the total power of the light elements was arrayed against the total and oppositely directed power of the heavy elements. If lightness predominated, upward motion would result; if heaviness, downward motion would follow. The light and heavy elements in mixed bodies were conceived as if they were composed of parts or degrees. A summation of the parts would reveal the predominance of heavy or light motive qualities and thus determine the direction of natural motion. The greater the ratio of heavy to light parts, the greater the downward speed; similarly, upward speed increased as the ratio of light to heavy elements increased.

From this interpretation it was an easy step to the concept of internal resistance. Because by their very natures heavy and light elements move in contrary directions, and because the practice had developed of assigning degrees to each of the elements in a compound, someone apparently took a further step and conceived heaviness and lightness as oppositely acting forces, or qualities, within the same mixed body. The quality – light or heavy – with the greatest number of degrees was designated the motive force, and its opposite functioned as resistance. If, now, two mixed bodies were compared, such that in one heaviness exceeded lightness by eight to three and in the other it exceeded it by eight to five, one could reasonably assume that in the same external medium, the body with the fewer degrees of lightness would fall with the greater velocity. This was explained by the fact that the quicker moving body had less lightness, or internal resistance. If both bodies had equal degrees of lightness and their downward speeds differed, it followed that the quicker descending body had more degrees of heaviness. Generally, in a falling body, heaviness was identified as the motive force and lightness as the resistance; in a rising body, lightness operated as the motive force and heaviness as the resistance.

Because all potentially observable physical bodies in the sublunar region were mixed bodies, internal resistance could be employed to explain all natural terrestrial motion. It was most useful, however, for justifying motion in a hypothetical void, because the accepted preconditions of motion were now present, namely a motive force and a resistance. In the absence of an external resistant medium, as in a vacuum, a body's internal resistance would prevent instantaneous speed. Because every mixed body contained within itself its own motive force and resistance, it was movable in void spaces.

But what about pure elemental bodies? Although, as we have already noted, such bodies were not found in nature, scholastics still considered whether these hypothetical bodies could move in a vacuum. Given the circumstances, the conclusion was obvious: they could not. A pure elemental body lacked internal resistance in the sense described for mixed bodies. And because there could be no external resistance in a void, no ratio of motive force to resistance was possible. It followed that the motion of an elemental body in a vacuum would be of infinite speed. Pure, unmixed elemental bodies, such as air, water, and earth, could fall only with finite speeds in material media (an absolutely light element, fire, could not possibly fall through any medium). Despite an occasional attempt to conjecture the means by which an elemental body could move with natural motion in a vacuum, such motion was generally considered dynamically unfeasible.

Within the context of medieval physics, internal resistance seemed the most reasonable way of justifying natural motion in a void for mixed bodies. Once this was established, an interesting and significant result was derived. Thomas Bradwardine, Albert of Saxony, and others concluded that two homogeneous bodies of different size and weight would fall in a void with equal speed. From the standpoint of Aristotelian physics, where speed is proportional to heaviness or absolute weight, so that the heavier a body the greater its velocity, this was a startling conclusion, a conclusion made possible by the assumption of material homogeneity. Because every equal unit of matter in every homogeneous mixed body is identical, every unit of matter must have the same ratio of heavy to light elements, that is, the same ratio of motive force to internal resistance, $F/R$. Although one body may contain more equal homogeneous units of matter than another and therefore be larger and weigh more, two such bodies will, nonetheless, fall with equal speeds; speed was held to be governed solely by an intensive factor, in this case the ratio of force to resistance per unit of matter, rather than an extensive factor such as total weight, as Aristotle had proposed.

Confronting the same problem more than two centuries later, Galileo employed a similar approach (in his *On Motion* [*De motu*], written around 1590) in rejecting Aristotle's explanation of natural fall. Instead of a ratio of force to internal resistance per unit of matter, Galileo relied on weight per unit volume, or specific weight. He argued that homogeneous bodies of unequal size, and therefore unequal weight, would fall with equal speeds in plenum and void, though their respective speeds in the latter would be greater than in the former. He was led to this conclusion by seizing upon effective weight, rather than gross weight, as the ultimate determinant of velocity. For Galileo, effective weight was equal to the difference in the specific weight of a body and the medium through which it fell. Hence it was actually a difference in specific weights that determined velocities. The velocity of a falling body may be represented as $V \propto$ specific weight of body minus specific weight of medium; the velocity of a rising body as $V \propto$ specific weight of medium minus specific weight of body. In a void, where the specific weight of the medium is zero, a body would fall with a velocity directly proportional to its specific weight, or weight per unit volume. Obviously, if the specific weights of two unequal bodies are equal, they will fall with equal speed in the same medium or in the void. Galileo extended the scope of this law in his most famous work, Discourses on Two New Sciences (Discorsi e Dimostrazioni Matematiche, intorno à due nuove scienze) (1638), where he declared that all bodies of whatever size and material composition would fall with equal speed in a vacuum, a generalization destined to become an integral part of Newtonian physics.

The similarity of approach and the nearly identical conclusions reached by Galileo and his medieval predecessors are striking, but perhaps only coincidental. Although Galileo may have been aware of medieval discussions, no specific evidence exists to substantiate this possibility. Indeed, where medieval Aristotelians explained direction of motion in terms of absolute lightness and heaviness functioning as motive qualities, Galileo relied on the relation between weight of body and medium. No longer was it necessary to distinguish the behavior of simple or pure elemental bodies from that of mixed bodies. The widely accepted medieval view that only certain bodies (mixed) could fall with finite speed in hypothetical void space, and that all others (pure elemental bodies) could not, became utterly meaningless in the physics of Galileo's On Motion. By accepting the fundamental concept of specific weight, Galileo treated all bodies alike, regardless of composition, and concluded that all bodies could fall or move in void and plenum. With Galileo, homogeneous Archimedean magnitudes replaced the elemental and heavy mixed bodies of the later Middle Ages. The inappropriateness of this dichotomy for Galileo's analysis of motion and his rejection of absolute heaviness and lightness rendered the concept of internal resistance meaningless. Internal resistance, which had been invoked in the Middle Ages to permit an explanation in dynamic terms for finite motion in the void, depended upon the contrary tendencies of elements distinguishable as light and heavy in a mixed body. In downward motion, heavy and light functioned as motive force and internal resistance, respectively; in upward motion their roles were reversed. Galileo, however, required neither internal nor external resistance for the production of finite speed in a void, where the speed of a falling body would be directly proportional to its specific weight. Force and the external resistance of a medium, where the latter was a factor, were measured objectively by specific weight. Although specific weight proved inadequate as a mode of explanation for natural fall, its use in the late sixteenth century by Galileo, and somewhat earlier in the century by Giovanni Battista Benedetti, represented an improvement over the vague and ill-defined notions of force and resistance that prevailed in the Middle Ages. But the medieval theories were themselves important departures from the received views that had entered Western Europe. They were efforts to cope with previously neglected aspects of Aristotelian physics and were interesting attempts to arrive at more generalized and comprehensive explanations.

Galileo's more consistent and simplified description of fall in void and plenum must not, however, obscure the historical fact that he inherited the very idea of the intelligibility of finite motion in a void from a tradition that is directly traceable to the Latin Middle Ages, to Avempace, and beyond to Philoponus. Indeed, Galileo indirectly acknowledged as much. It was from this anti-Aristotelian tradition that he derived the idea that a resistant medium is merely a retarding factor in the fall of bodies, whose real natural motions occur only in a vacuum, albeit a hypothetical one.

By taking his conclusion that homogeneous bodies fall with equal speeds in a vacuum and extending its scope to all bodies of any whatever composition, Galileo eventually made a new law of physics. His medieval predecessors arrived at the same conclusion for homogenous bodies, but extended it no further. Did this important consequence have any significant repercussions in the consideration of motion in medieval natural philosophy? Like other departures from Aristotle, it appears to have had little overall effect. It never occurred to medieval natural philosophers to investigate its possible implications for other aspects of Aristotelian natural philosophy, a reaction that was all too typical.

*Violent motion in a vacuum and impetus theory.* By contrast with the lengthy and rather numerous discussions about the possibility of natural motion in a vacuum, the possibility of violent motion was barely considered. The problem was formidable because neither of the two essentials in violent motion, namely, external motive force and external resistance, were present in a vacuum. In the absence of a physical medium such as air or water, no external motive force or resistance could be invoked, as in Aristotle's explanation of violent motion. Lightness and heaviness, which functioned as internal force and resistance in mixed bodies undergoing natural motion, were of little use in accounting for violent motion. A mixed body in which heaviness predominated must, by definition, be moved upward (away from its natural place) in violent motion, or horizontally, so that its predominant heaviness could not operate as a motive force. Other than denying the possibility of violent motion in a vacuum, the only reasonable response consistent with late medieval physical principles is embodied in a statement by Nicholas

Bonetus (d.1343?), who related that "in a violent motion some non-permanent and transient form is impressed in the mobile so that motion in a void is possible as long as this form endures; but when it disappears the motion ceases."[1]

Bonetus is here referring to an impressed force, one of the most important physical concepts used in the late Middle Ages and one that had emerged centuries earlier by way of disagreement with Aristotle's explanation of violent motion. Already in late antiquity, Aristotle's identification of the external air as the continuing motive force in violent motion came under attack. John Philoponus observed that if air in direct contact with an object could cause and maintain that object's motion for a time, as Aristotle supposed, then it ought to be possible, for example, to put a stone into motion by merely agitating the air behind it. Because this was contrary to experience, Philoponus rejected air as a motive force and assumed instead that an incorporeal motive force, imparted by an initial mover to a stone, or projectile, was the cause that enabled the stone to continue its movement. With the impressed force acting as motive power and the stone, or object, functioning as resistance, the requirements for violent motion were met. The surrounding air contributed little or nothing to this process. Indeed, it was an obstacle to continuing motion. Philoponus concluded that violent motion would occur more readily in a vacuum than in a plenum, because no external resistance could impede the action of the impressed force.

Further elaboration of Philoponus's explanation was made by Islamic authors, to whom the impressed force was known as *mail* (inclination, or tendency). One of the major Islamic proponents of the *mail* theory was Avicenna, who conceived of it as an instrument of the original motive force. *Mail* was capable of continuing the action in a body after the original motive force was no longer operative. Avicenna distinguished three types of *mail*: psychic, natural, and violent. Leaving aside the first, which is not relevant to our discussion, natural and violent *mail* were intended to furnish causal explanations for the two corresponding types of motion differentiated by Aristotle. According to Avicenna, a body was capable of receiving violent *mail* in proportion to its weight. This explained, for example, why a small lead ball can be thrown a greater distance than a piece of light wood, or a feather. Ontologically, Avicenna conceived of *mail* as a permanent quality, which, in the absence of external resistances, would endure in a body indefinitely. He therefore concluded that if a body were moved violently in a void of indefinite extent, its motion would be of indefinite duration, because there would be no reason for it to come to rest, a conclusion that Aristotle had also reached (without invoking impressed forces) and for which reason, among others, he had rejected the existence of void space. Because Avicenna found no evidence that motions of this kind existed, he also denied the existence of void space.

In the next century, Abu'1 Barakat (d. ca. 1164) proposed a different kind of *mail*, one that was self-dissipating, the type described by Nicholas Bonetus much later. Thus even in a void, a body in violent motion would eventually cease movement because of the natural and inevitable exhaustion of its impressed force, a consequence that could not be used as a serious argument against the existence of void. Arab authors had thus described two different forms of impressed force, or impetus. One was destructible only by external forces and obstacles and was otherwise permanent; the other was transient and self-dissipating and would gradually disappear over time, even in the absence of external forces. Counterparts to these two types of impressed force eventually appeared in the Latin West. Whether they were transmitted via Latin translations of Arabic works or were independently developed in the Latin West is uncertain.

The theory of impressed force was already known in the thirteenth century, because a few Latin authors, such as Roger Bacon and Thomas Aquinas, rejected the idea that an incorporeal impressed force could account for the continued violent motion of a body. It was not until the fourteenth century, however, that forms of impressed force theory became popular, especially at Paris. As early as 1323, Franciscus de Marchia proposed one version in which the incorporeal impressed force, or *virtus derelicta* (the force left behind), as he called it, was a naturally self-dissipating temporary force capable of moving a body contrary to its natural inclination. In this process, air continued to play a subsidiary role. For when the body was set in motion, Franciscus believed that the surrounding air also received an impressed force that enabled it to assist the body's motion.

The best-formed theory was presented by John Buridan, who is perhaps also responsible for the introduction of *impetus* as a technical term for impressed incorporeal force. Buridan conceived of impetus as a motive force transmitted from the initial mover to the body set in motion. The speed and the quantity of matter of a body were taken as measures of the strength of the impetus producing the motion. On the correct assumption that there is more matter in a heavy, dense body than in a lighter, rarer body of the same volume and shape, Buridan explained that if a piece of iron and a piece of wood of identical shape and volume were moved with the same speed, the iron would traverse a greater distance, because its greater quantity of matter could receive more impetus and retain it longer against external resistances. Thus it was that Buridan seized upon *quantity of matter* and *speed* as means of determining the measure of impetus. These were the same quantities that defined momentum in Newtonian physics, although there momentum was usually conceived of as a quantity of motion, or a measure of the effect of a body's motion, whereas impetus was a cause of motion. Indeed, impetus was envisioned as an internalization of Aristotle's external motive force. It seemed a better way of adhering to Aristotle's own dictum that everything that is moved is moved by another.

Like Avicenna, Buridan ascribed to impetus a quality of permanence and assumed that it would last indefinitely, unless diminished or corrupted by external resistance. Once a mover imparted impetus to a body and the latter moved off and lost contact with the original motive force, no additional impetus could be produced in the absence of any identifiable cause. The initial quantity of impetus would remain constant unless corrupted by external resistances acting on the body, or by the natural tendency of the body to move to its natural place. Buridan implied that if all resistances to motion could somehow be removed, a body set in motion would move indefinitely, presumably, in a straight line at uniform speed. There would be no reason for it to change direction or alter its initial speed because not even its inclination to fall to its natural place would be operative as long as all impediments to its forced motion have been removed. Indeed, while the impetus was producing a violent motion away from its natural place, the body's inclination to fall toward its natural place would, presumably, be inoperative. Buridan failed to elaborate this potential inertial consequence of impetus theory, perhaps because the very idea of an indefinite rectilinear motion under the ideal conditions just described would have appeared absurd in a finite Aristotelian world. If he thought that such an indefinite rectilinear motion was really possible, he would probably have devised a mechanism for stopping it. In truth, Buridan avoided dilemmas of this kind by denying the possibility of finite, successive motion in a void. As a consequence of the Condemnation of 1277, however, he conceded that God could supernaturally produce such motions in a vacuum. If Buridan had adopted a nonpermanent, or self-dissipating, variety of impetus, as described in the earlier statement from Nicholas Bonetus and accepted for a time by Galileo, he might have accepted motion in a hypothetical void. With a nonpermanent impetus, motion in a void could only have been of finite duration.

Although the concept of an indefinite, uniform, rectilinear motion – an essential ingredient in the principle of inertia – was incompatible with medieval physics, Buridan's permanent impetus incorporated characteristics and properties from which such a motion could be derived. Before Newton conceived of inertia as an internal force that enabled bodies to resist changes in their states of rest or uniform rectilinear motion (the idea that rest and uniform rectilinear motion are identical states of a body was never suggested in the Middle Ages, where rest and motion were considered contrary attributes, or states), he had thought of inertia much as Buridan thought of impetus, namely, as an internal force that, in the absence of external forces or resistances, would cause indefinite rectilinear motion.

In a manner similar to earlier Islamic *mail* theorists, Buridan also applied impetus to explain the acceleration of falling bodies. Throughout the history of physics, up to and including Galileo, the problem of fall was treated in a twofold manner: the first way was to explain the cause of fall in general without regard to its acknowledged acceleration; the second was concerned with its acceleration. We saw earlier that Aristotle had suggested the generator of a thing as the cause of its natural fall but, in his actual discussions, had emphasized weight as the determinant of a heavy body's uniform downward speed. The body's acceleration was virtually ignored. In the medieval Latin West, some authors identified a body's substantial form as the cause of its fall, whereas others, especially in the fourteenth century, considered the heaviness or weight of a body as the primary cause of fall. To account for acceleration, a second, quite separate, cause was sometimes added.

Buridan approached the problem in this way. Because the weight of a body remained constant as it fell, he identified the body's heaviness, or gravity (*gravitas*), as the cause of its natural uniform fall. After eliminating a few commonly discussed possible causes of acceleration (such as proximity to natural place, rarefaction of the air from the heat produced by the falling body, and the lessening of air resistance as the body descended), Buridan explained acceleration by accumulated increments of impetus. The heaviness of a body not only initiated its downward fall, but also produced an increment of impetus, or "accidental heaviness," as it was sometimes called, during each successive moment of fall. The successive and cumulative increments of impetus then generate successive and cumulative increments of velocity, thus producing a continuously accelerated motion.

Three elements are distinguishable in the process of fall: ( 1) heaviness of the body, $W$; (2) impetus, $I$; and (3) velocity, $V$. Initially, at the end of the first temporal instant, $\Delta t$, the heaviness, or weight, $W$, produces an original velocity, $V$. During the same time interval, the heaviness of the body, which remains constant, simultaneously produces a quantity of impetus, $I$, which will be operative during the second instant of time and produce an increment of velocity, $\Delta V$. Thus at the end of the second interval of time, $2\Delta t$, the body's heaviness and the increment of impetus, $W + I$, increase the body's speed to $V + \Delta V$. During the second time interval, $2\Delta t$, a second increment of impetus is generated and added to the first. Therefore, in time interval $3\Delta t$, $W + 2I$ will produce a speed of $V + 2\Delta V$. In the fourth interval, $W + 3I$ will increase the speed to $V + 3\Delta V$, and so on. Buridan's explanation is squarely in the Aristotelian tradition, because force is always proportional to velocity and not to acceleration as would be the case in Newtonian physics. This is obvious because every increment of velocity is preceded by a proportional increment of impetus. Thus if after force $W + 3I$ produced $V + 3\Delta V$ no additional increments of impetus were added, the speed would remain constant at $V + 3\Delta V$ and remain proportional to a now constant force of $W + I$. Only if weight were taken as a constant motive force producing increments of velocity directly, rather than increments of impetus, could it be argued that Buridan had arrived at something similar to the concept that force is proportional to acceleration. But there is little warrant for this interpretation – the body's weight must first produce an increment of impetus before a proportional increment of velocity can be generated. The connection between weight as a constant motive force and increase of velocity is at best indirect.

Despite some opposition, impressed force theories exercised a continuing influence into the sixteenth century, when Galileo himself became an enthusiastic proponent during his early career at the University of Pisa. In his *On Motion*, which remained unpublished during his lifetime, Galileo sought to explain the forced upward motion and subsequent downward acceleration of heavy bodies. As the basis of his explanation, he adopted the idea of a residual force, which he derived from Hipparchus (d. after 127 B.C.), whose views were described in Simplicius's *Commentary on Aristotle's On the Heavens*, a treatise widely known in the Middle Ages. To this, Galileo added the mechanism of a self-dissipating incorporeal impressed force, or impetus, which he probably derived, ultimately, from medieval sources. Initially, the mover imparts an impressed force to a stone that is hurled aloft. As the force diminishes, the body gradually decreases its upward speed until the impressed force is counterbalanced by the downward thrust of the stone's weight, at which moment the stone begins to fall, slowly at first, and then more quickly, as the impressed force diminishes and gradually dissipates itself. Acceleration results as the difference between the weight of the stone and the power of the diminishing impressed force continually increases. Thus, on the downward leg of the motion, the impressed force actually functions as a resistance. If the body fell a sufficiently long distance, all of its impressed force would vanish, at which point it would fall with uniform speed. Eventually, Galileo abandoned the concept of a self-dissipating impressed force and explained accelerated fall by an impetus that is conserved and cumulative, an explanation that differed little from Buridan's.

## The kinematics of motion

If medieval natural philosophers could be said to have departed dramatically from Aristotle with respect to the dynamics, or causes, of motion, by largely abandoning or altering many of his ideas, they made their contributions to the kinematics of terrestrial motion as radical extensions of ideas that Aristotle had barely hinted at or, as is more likely, had never thought of. Nevertheless, it was Aristotle who first raised the issue, which eventually produced significant, though unintended, results.

*Motion as the quantification of a quality: The intension and remission of forms.* In chapter 8 of his *Categories*, Aristotle explained that variation in degrees is found not only among sensible qualities but also in abstract qualities, such as justice and health (one person might be healthier or more just than another). He suggested two alternatives, without choosing between them: either the quality itself varies, or – and this was the opinion usually ascribed to Aristotle during the late Middle Ages – the subject participates to a greater or lesser degree in an invariant quality or form.

With Aristotle's *Categories* as background, the problem of qualitative variation acquired greater significance as the result of a theological question that was posed in the twelfth century by Peter Lombard in the first book of his famous *Sentences* (distinction 17). Here Peter asked "whether it ought to be conceded that the Holy Spirit could be increased in man, [that is,] whether more or less [of it] could be had or given." Peter insisted that charity, or grace, which is caused by the Holy Spirit, could not vary in humans because this would imply a change in the Holy Spirit. Theologians generally assumed that charity was a constant spiritual entity that individuals could possess more of or less of by greater or lesser participation in it, an opinion analogous to Aristotle's claim that although justice remained constant, it varied in people by virtue of their greater or lesser participation in it.

This theory may be characterized as the popular doctrine of participation, which was adopted by Thomas Aquinas. It was a second theory, however, that would triumph. The quality itself, not the degree of participation, was taken as the variable. After numerous interpretations of this second theory were proposed, a widely accepted version, associated with the name of John Duns Scotus (ca. 1265–1308), emerged in the fourteenth century. A quality could be increased by the formal addition of new, real, and distinct similar parts to an already existing form or quality. The pre-existing parts would join with the new parts to constitute a unified form of a definite intensity. Similarly, qualities were diminishable by losing distinct parts. In this approach, every quality was considered augmentable or diminishable, as if it were an extended magnitude, or a weight. Just as one weight added to another weight produced a new and increased weight, so the addition of one qualitative part to another would augment the intensity of any variable quality.

Although qualities are not like extensive magnitudes, the mathematical treatment of qualitative intensities in the fourteenth century, known as the "intension and remission of forms" or "qualities," was influenced by the conviction that such operations were conceptually meaningful. Scholastic natural philosophers who treated this subject became interested primarily in the mathematical aspects of qualitative change and less interested in the theological and ontological aspects that had been prominent earlier.

A significant development occurred around 1330, when the process of intension and remission of forms was linked analogically with motion, so that velocity, which was a successive and nonpermanent entity, came to be treated as if it were a permanent but variable quality, like color or taste. Thus, just as a succession of forms of different intensities was assumed to explain continuous increase or decrease of a quality's intensity, so was the succession of new positions acquired by a motion viewed as a succession of forms representing new degrees of that motion's intensity.

During the early fourteenth century, at Merton College (Oxford University), a group of English scholars – among whom the more prominent were William Heytesbury, John Dumbleton, and Richard Swineshead – came to treat variations in velocity, or local motion, in the same manner as variations in the intensity of a quality. The intensity of a velocity increased with speed, no less than the redness of an apple increased with ripening. For the next 300 years, from the fourteenth through sixteenth centuries, the analogy between variable qualities and velocity was a permanent feature of treatises on intension and remission of forms and qualities. Although the study of qualitative variations persisted for a long time, the developments that emerged and proved significant in the history of physics were all formulated in the fourteenth century at the Universities of Oxford and Paris, but only after the initial theological and metaphysical context within which these problems had originally been discussed were abandoned or ignored.

The medieval contribution centered on original and correct definitions of uniform speed and uniformly accelerated motion, definitions employed by Galileo and on which he did not improve. At Merton College, and elsewhere, uniform motion was defined as the traversal of equal distances in *any* (or *all*) equal time intervals. Like Galileo, some medieval authors cogently added the word "any" in order to avoid the possibility of equal distances being traversed in equal times by nonuniform velocities. In generalizing that equal distances are traversed in any and all time intervals, however small or large, their definition guaranteed uniformity of motion.

Extending the definition of uniform motion to the simplest kind of variable speed, the Mertonians arrived at a precise definition of uniform acceleration as a motion in which an equal increment of velocity is acquired in each of any equal intervals of time, however large or small.

They also sought to define the difficult notion of an instantaneous velocity. Lacking the fundamental concept of a limit of a ratio, which was only developed centuries later in the calculus, they defined uniform acceleration in terms of uniform velocity. Thus uniform acceleration was expressed as the distance that would be traversed by a moving point or body if that point or body were moved uniformly over a period of time with the same speed it possessed at the instant in question. Although the definition is hopelessly circular, in that it defines "instantaneous velocity" by a uniform speed equal to the very instantaneous velocity that is to be defined, the Mertonians merit praise for recognizing the need for such a concept. Galileo employed it in the same form. Not only did these scholars cope with instantaneous velocity directly, if inadequately, by definition, but they also approached it indirectly, through their definitions of uniform and uniformly accelerated motion, where velocities in infinitesimally small time intervals are clearly implied.

By an admirable and ingenious use of these definitions, the Merton scholars derived what is known as the mean speed theorem, probably the most outstanding single medieval contribution to the history of mathematical physics. In symbols, it may be expressed as $S = \frac{1}{2}V_f t$, where $S$ is the distance traversed, $V_f$ is the final velocity, and $t$ is the time of acceleration. Because the velocity is assumed to be uniformly accelerated, $V_f = at$, where $a$ is uniform acceleration, we get, by substitution $s = \frac{1}{2}at^2$, the usual formulation for distance traversed by a uniformly accelerated motion starting from rest. When, instead of starting from rest, the uniform acceleration commences from some particular velocity, $V_0$, a case frequently discussed, the medieval version is representable as $S = [V_0 + (V_f - V_0)/2]t$, or simply $S = V_0 t + \frac{1}{2}at^2$, since $V_f - V_0 = at$.

The mathematical expressions and symbols used in the preceding paragraph were nonexistent in the Middle Ages. The concise symbolic formulations used here were expressed rhetorically in a manner that might seem cumbersome and prolix, and perhaps even be incomprehensible, to present-day readers. Here is a version from William Heytesbury's *Rules for Solving Sophisms* (*Regule solvendi sophismata*), composed around 1335:

> For whether it [i.e., a latitude or increment of velocity] commences from zero degree or from some [finite] degree, every latitude, as long as it is terminated at some finite degree, and as long as it is acquired or lost uniformly, will correspond to its mean degree [of velocity]. Thus the moving body, acquiring or losing this latitude uniformly during some assigned period of time, will traverse a distance exactly equal to what it would traverse in an equal period of time if it were moved uniformly at its mean degree [of velocity].[2]

A modern English verbalization of this important theorem would probably seem no less difficult or verbose than Heytesbury's medieval version. For example, we might explain it as follows:

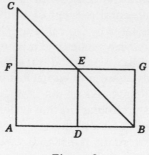

*Figure 2*

a body, or point, commencing with uniform acceleration from rest, or from some particular velocity, would traverse a certain distance in a certain time. If the same body were to move during the same interval of time with a uniform velocity equal to the instantaneous speed acquired at the middle instant of its uniform acceleration, it would traverse an equal distance.

A uniformly accelerated motion is thus equated with a uniform motion, making it possible to express the distance traversed by the former in terms of the distance traversed by the latter. Numerous arithmetic and geometric proofs of this vital theorem were proposed during the fourteenth and fifteenth centuries. Of these, the best known is a geometric proof by Nicole Oresme, formulated around 1350 in a work titled *On the Configurations of Qualities and Motions*, easily the most original and comprehensive extant treatment of the intension and remission of qualities.

In Figure 2, let line *AB* represent time and let the perpendiculars erected on *AB* represent the velocity of a body, Z, beginning from rest at *B* and increasing uniformly to a certain maximum velocity at *AC*. The totality of velocity intensities contained in triangle *CBA* was conceived as representing the total distance traversed by Z in moving from *B* to *C* along line *BC* in the total time *AB*. Let line *DE* represent the instantaneous velocity that Z acquires at the middle instant of the time as measured along *AB*. If Z were now moved uniformly with whatever velocity it had at *DE*, the total distance it will traverse in moving from *G* to *F* along line *GF* in time *AB* is represented by rectangle *AFGB*. If it can be shown that the area of triangle *CBA* equals the area of rectangle *AFGB*, it will have been demonstrated that a body accelerated uniformly from rest would traverse the same distance as a body moving during the same time interval at a uniform speed equal to that of the middle instant of the uniformly accelerated motion. That is, $S = \frac{1}{2}V_f t$, the distance traversed by Z with uniform motion, equals $\frac{1}{2}at^2$, the distance traversed by Z when it is uniformly accelerated. That the two areas are equal is demonstrated as follows: because $\angle BEG = \angle CEF$ (vertical angles are equal),

∠BGE = ∠CFE (both are right angles), and *GE = EF* (line *DE* bisects line *GF*), triangles *EFC* and *EGB* are equal (by Euclid's *Elements*, book I, proposition 26). When to each of these triangles is added area *BEFA* to form triangle *CBA* and rectangle *AFGB*, it is immediately obvious that the areas of triangle *CBA* and rectangle *AFGB* are equal.

Oresme's geometrical proof, as well as numerous arithmetical proofs of the mean speed theorem, was widely disseminated in Europe during the fourteenth and fifteenth centuries, and was especially popular in Italy. Through printed editions of the late fifteenth and early sixteenth centuries, it is possible that Galileo learned about this by then well-known proof. He made the mean speed theorem the first proposition of the third day of his *Discourses on Two New Sciences*, where it served as the foundation of the new science of motion. Not only is Galileo's proof strikingly similar to Oresme's, but the accompanying geometric figure is virtually identical, despite a ninety-degree reorientation, a reorientation that had already been made by some medieval authors.

If the major theorems and corollaries attributed to Galileo were already enunciated in the Middle Ages, in what sense, if at all, is it arguable that Galileo founded the modern science of mechanics? The rectification of the historical record has not diminished Galileo's stature, nor has it deprived him of the right to be honored as the founder of mechanics. Although Galileo had been anticipated by some of the medieval contributions described here, his originality derives from an exceptional ability to seize upon and extract what was directly relevant to a mathematical and kinematic description of motion from the diffuse and unfocused medieval doctrine of intension and remission of forms and qualities. The many ingenious conclusions and theorems derived in treatises on the intension and remission of qualities and velocities during the fourteenth to sixteenth centuries were little more than intellectual exercises reflecting the subtle imagination and logical acumen of scholastic thinkers. With one minor exception, these scholars were content to treat velocities as variable, intensive qualities divorced from the motion of real bodies. Oresme, for example, characterized geometrical representations of quality variations as fictions of the mind without relevance to nature. Not until the sixteenth century did an obscure scholastic author think of applying the mean speed theorem to naturally falling bodies. Sometime around 1545, Domingo Soto declared, in his *Questions on the Physics of Aristotle*, that a body that falls through a homogeneous medium from some height increases its motion "uniformly difformly," that is, falls with a uniform acceleration. He invoked the famous example, first used by scholars at Merton College, that a body accelerating uniformly from zero degree of speed, or rest, to a speed of 8, will traverse the same distance as a body moving through the same time with a uniform speed of 4, which is the mean speed of the uniformly accelerating body. Although Soto believed that uniformly accelerated motion is illustrated in naturally falling bodies, he did nothing more.

By contrast, Galileo pursued the matter with great vigor and genius. He brought together all the significant concepts, definitions, theorems, and corollaries on motion and arranged them into a logical and ordered whole, which he then applied to the motion of real bodies. Uniform acceleration was no longer a mere hypothetical definition, but a true description of the way bodies fall in nature, as exemplified by Galileo's famous inclined-plane experiment. Galileo constructed a new science of mechanics and thereby laid the foundations of modern physics. What he wrought became a vital part of Newtonian science. This achievement alone suffices to include Galileo among that small group of extraordinary scientists who, from time to time, profoundly alter the character and direction of science.

Aristotle's ideas about motion in both plenum and vacuum provided the basis for some of the most dramatic departures from his natural philosophy that occurred with respect to terrestrial phenomena. But what about the celestial region? What sort of departures could occur in a region of the world in which change was presumed absent?

## THE CELESTIAL REGION

Two major cosmological systems entered Western Europe in the twelfth century: Aristotle's and Ptolemy's. The relevant ideas of each was represented by more than one work. Aristotle's cosmological thought is largely found in his *On the Heavens*, *Physics*, *Metaphysics*, and *Meteorology*, whereas Ptolemy's cosmological ideas appear primarily in his *Hypotheses of the Planets*, which was known only indirectly during the late Middle Ages. Two other works by Ptolemy also play a lesser role, namely, the *Almagest*, a technical, geometrical treatise on astronomy, and the *Tetrabiblos*, an astrological work. Aristotle and Ptolemy shared a few basic concepts. They both believed that each of the seven planets was embedded in its own ethereal sphere and was carried around by it. They further assumed that the fixed stars were located on a single sphere that surrounded and encompassed all the planetary orbs. Finally, both agreed that each planetary sphere consisted of a plurality of subspheres that were needed to account for the resultant motion and position of the planet that was being carried around. In effect, Aristotle and Ptolemy constructed systems with a multiplicity of nested orbs, or, better, rings or shells – as many as fifty-five in Aristotle's system and, depending on which of two versions he described, thirty-four, or forty-one, or even twenty-nine in Ptolemy's.

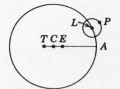

*Figure 3    Epicycle on eccentric circle. The distance of the planet (P) from the earth varies with its position on the epicycle. "The motion of the center (L) of the epicycle that carried the planet (P) was regarded as uniform not with respect to the center (C) of the deferent [i.e., the circular path traced by (L)], or of the earth (T), but with respect to another point (E), called the 'equant,' i.e., ∠ LEA was considered as increasing uniformly. By properly locating the points E, C, and T, and determining the ratio of the diameters of the epicycle and its deferent, and choosing proper directions, velocities, and inclinations for the various circles, the apparent irregularities were accounted for." (The figure and accompanying legend are reprinted by permission of the Harvard University Press from Morris R. Cohen and Israel Drabkin,* A Source Book in Greek Science, *Harvard University Press, 1958, p. 129. I have added the bracketed description.)*

### The three-orb compromise

Despite their similarities, the systems of Aristotle and Ptolemy differed radically. Aristotle's spheres were concentric with respect to the earth, whereas Ptolemy's were eccentric and epicyclic. Concentric orbs, which had the earth as their center, could not account for observed variations in the distances of the planets. In his *Almagest*, Ptolemy employs eccentric and epicyclic circles to account for such variations (see Figure 3) and thus corrects the fundamental deficiency in the system of concentric orbs, of which Aristotle's was one of the last of any consequence. Ptolemy's geometric methods in the *Almagest* were intended to account only for planetary positions and were not regarded as a true representation of the physical world. It was in his *Hypotheses of the Planets* that Ptolemy sought to depict the physical world and the relationships of its celestial orbs. Although medieval natural philosophers knew Ptolemy's *Hypotheses of the Planets* only indirectly, they adopted a compromise from it that was easily made compatible with Aristotle's concentric system. The compromise hinged on a distinction between the concept of a "total orb" (*orbis totalis*) and a "partial orb" (*orbis partialis*). The former is a concentric orb whose center is the earth's center, or the center of the world, whereas the latter is an eccentric orb, the center of which lies outside the center

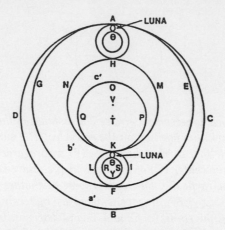

*Figure 4*  Let T *be the center of the earth and world and also the center of the lunar orb. The "total orb" of the Moon lies between the convex circumference* ADBC *and the concave circumference* OQKP, *which are both concentric to* T. *Between these two circumferences, three "partial" eccentric orbs are distinguished* (a', b', *and* c') *by assigning another center,* V, *toward the Moon's aux, or apogee. Around* V *as center are two circumferences,* AGFE *and* HNKM, *which enclose the lunar deferent and form the eccentric deferent, or orb* b'. *Surrounding the eccentric orb is the outermost orb,* a', *lying between surfaces* ADBC *and* AGFE; *and surrounded by the eccentric orb is the innermost orb,* c', *lying between the concave surface* HNKM *and the convex surface* OQKP. *Inside the concavity of the eccentric deferent lies a spherical epicycle, which was conceived either as a solid globe that lacks a concave surface, or as a ring with two surfaces, one convex* (KLFI), *the other concave* (RYSΘ). *This representation of the Moon's concentric, eccentric and epicyclic orbs is described in Roger Bacon's* Opus tertium. *The diagram and Bacon's text appear in Pierre Duhem, ed.,* Un fragment inédit de l' "Opus tertium" de Roger Bacon précédé d'une étude sur ce fragment *(Florence, 1909), 129.*

of the world. Every total orb consists of at least three partial orbs, for which reason we may describe it as the "three-orb system" (see Figure 4, where the Moon's three orbs are depicted). The convex and concave surfaces of a total concentric orb, or ring, have the earth's center, or the center of the world, as their center. The three partial orbs lie inside these two concentric surfaces. Within the middle partial orb, called the eccentric deferent, lies an epicycle in which the planet is located.

In this manner were the two cosmologies of Aristotle and Ptolemy joined. The concept of the total orb allowed for the preservation of Aristotle's concentric cosmology, whereas the concept of partial eccentric orbs permitted the variation of planetary distances without which Aristotle's cosmology would have been unacceptable. The inclusion of eccentric orbs within Aristotelian concentric spheres meant that Aristotelian cosmology had to accept some unpleasant "realities." Although each total, concentric orb rotated around the physical earth as its center, the eccentric orbs within it did not have the earth as their center, but rotated around a geometric point that was eccentric to the earth. The assumption of eccentric orbs – or any orbs whatever – moving around points that lay outside the center of the world was contrary to Aristotelian cosmology and physics. In this instance, however, the needs of astronomy, in the form of planetary variations in distance, had to take precedence over Aristotle's astronomically unacceptable cosmology. The options were obvious: either Aristotelians had to abandon the centrality of the earth and jeopardize the physics and cosmology that Aristotle had erected on the assumption that the earth lay at the center of all celestial orbs, or they had to retain a cosmology that was astronomically untenable. Medieval natural philosophers resolved the dilemma by adopting the three-orb compromise, thus making a significant departure from Aristotle's world view, but one that caused little concern. Because they were largely ignorant of technical astronomy, medieval natural philosophers generally disregarded the astronomical role played by the partial eccentric orbs that lay between the convex and concave surfaces of a concentric, or total, orb. Such difficulties were better left to astronomers. What mattered was that the foundations of Aristotelian cosmology were preserved, while honoring the basic tenets of Ptolemaic astronomy. The glaring fact that partial eccentric orbs did not have the earth as their physical center was simply ignored.

### The number of total orbs

Aristotle assigned no spheres beyond the fixed stars, which he regarded as the outermost orb of the cosmos. Subsequent developments in astronomy, and the needs of Christian theology, led natural philosophers to assign at least three more total spheres beyond the fixed stars (see Figure 5). Aristotle had assigned only the daily motion to the sphere of the fixed stars. In the course of later Greek and Islamic astronomy, two additional motions were assigned to it: a precession of the equinoxes, which was a west-to-east motion of the fixed stars that occurred at the rate of one degree in 100 years and produced a complete revolution of the starry sphere in 36,000 years, and a second motion, trepidation, an allegedly progressive and regressive motion of the stars proposed by Thabit ibn Qurra, a ninth-century Arab astronomer. On the Aristotelian principle that a single sphere was required for each distinct celestial motion, one

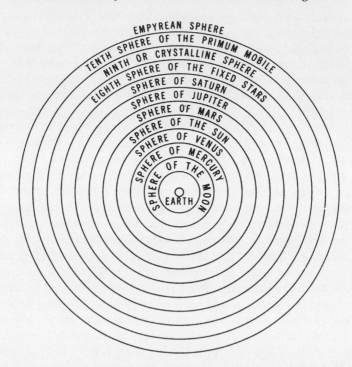

*Figure 5    A typical schematic representation of total spheres. Between 1200 and 1600, most diagrammatic depictions of the cosmos represented only total, concentric orbs, ignoring those that were partial and eccentric.*

additional sphere had to be assigned for precession and one for trepidation, for a total of ten orbs. Neither of these two spheres carried a celestial body. The ninth or tenth sphere, but sometimes both, was usually identified with the biblical waters above the firmament, the latter usually equated with the eighth sphere of the fixed stars. During the early Middle Ages, these waters came to be called *crystalline*, a term that was applied to fluid as well as to congealed waters, the latter in the form of ice or crystal. Supporters for each alternative could be found. For Saint Jerome (ca. 347–419?) and the Venerable Bede (672–735), the waters were hard and crystal-like, whereas for Saints Basil and Ambrose they were fluid and soft. The tenth sphere in Figure 5 was called the "first movable sphere" (*primum mobile*) because it was the first moving sphere enclosed by the outermost, immobile, empyrean heaven.

The orbs we have spoken of thus far were assumed to be moving with uniform, circular motion. All had some astronomical function. The empyrean heaven, or sphere, is a notable exception. Not only did it have no astronomical purpose, but it was conceived as immobile, thus differing from all other celestial orbs. It was a purely theological creation, a product of faith, not of science. However, despite its occasional identification with the heaven created on the first day, it had no biblical sanction and thus differed from the other two theological spheres, the firmament and the crystalline sphere. The empyrean emerged as a separate heavenly sphere only in the twelfth century, when theologians, such as Anselm of Laon, Peter Lombard, and Hugh of Saint Victor, described it as a place of dazzling luminosity, where God, angels, and the blessed dwelled through all eternity. Despite its perpetually radiant state, the empyrean heaven transmitted none of the light that filled it. As a sphere, it was transparent, invisible, and incorruptible. Nothing existed beyond its convex surface. As Campanus of Novara put it in the thirteenth century, "It is the common and most general 'place' for all things which have position, in that it contains everything and is itself contained by nothing."[3] With the all-encompassing, immobile empyrean sphere, we arrive at a total of eleven concentric spheres occupying and filling the cosmos.

### Celestial incorruptibility and change

The celestial spheres, as we saw, were assumed to be composed of an incorruptible ether whose properties contrasted radically with those of the incessantly changing four elements of the terrestrial region. In Aristotle's scheme, change was unrealizable in the heavens because of an absence of contrary qualities, so that no pair of such qualities could act on the same celestial body or in the same portion of celestial ether. But if the heavenly ether lacked contrary qualities, why did natural philosophers and astrologers attribute contrary qualities to celestial bodies? For example, why did they describe Saturn as cold and dry, Mars as hot and dry, the Moon as cold and wet, and so on for the other planets? If these qualities were really absent from celestial bodies, why did astrologers and natural philosophers speak as if they were present?

The standard medieval reply to such a question would have been that the celestial ether possesses these qualities only virtually (*virtualiter*), not formally (*formaliter*), or actually. That is, a celestial body is said to "possess" qualities like hotness or coldness only in the sense that it can somehow cause changes in the hotness or coldness in bodies below the Moon, even though that celestial body does not actually possess the qualities that it allegedly produces in terrestrial bodies. To say, for example, that Saturn is cold is not to say that the quality "coldness" actually inheres in Saturn's ethereal matter – indeed, it does not – but rather that Saturn has the capacity to produce the effect of coldness in the terrestrial region. Again, the Sun is not actually hot, but it has the capacity to cause hotness in objects in the terrestrial region.

The celestial existence of one pair of opposites – rarity and density – was, however, more than virtual. Medieval natural philosophers often assumed that some parts of the heavens were rarer or denser than others. Stars and planets were capable of reflecting light, and were therefore visible, because they were believed to be denser than the invisible celestial orbs that carried them. Scholastic authors were thus prepared to concede that the celestial ether was rarer in some places and denser in others. However, they would have denied that any particular part of the ether could vary in density or rarity. No planet, star, or orb could alter its density or rarity. Although parts of the ether were denser or rarer than other parts, so that the ether spread over the whole celestial region could not be homogeneous – contrary to Aristotle's belief – none of these parts ever altered its density or rarity. The ether as a whole remained in an invariant state. Thus even if it had been assumed that Saturn was actually, rather than only virtually, cold, and that the Sun was actually hot, rather than virtually hot, generation and corruption could not occur because Saturn did not possess the contrary hotness, and the Sun lacked the contrary coldness. Under these circumstances, hotness and coldness could be in the heavens simultaneously, but because they would never be together in the same body or part, they could not cause generation or corruption. Similarly, if density were to exist in one part of the heavens, or in one celestial body, and rarity were to exist simultaneously in another, they would not oppose each other. Thus rarity and density, hotness and coldness, and other pairs of contraries could exist in isolation in the heavens, as long as no pair was embodied in the same celestial subject. How an allegedly homogeneous ether could be differentiable into seven planets, each with different properties, and an innumerable host of fixed stars is a question that seems not to have occurred to medieval natural philosophers; or, if it did, it obviously caused little concern.

### The causes of celestial motion

The most frequently discussed theme in medieval cosmology was celestial motion. The questions were wide-ranging. Here I shall focus on the causes of celestial motion. What caused the invisible, celestial orbs to move around the sky with uniform motion? As we saw in chapter 4, Aristotle bequeathed two different explanations to account for celestial motions, one that relied on external causes and the other on internal causes. Although some poets and minstrels may have been captivated with Aristotle's idea that love makes the world go round (see chapter

4), medieval natural philosophers devised a more direct efficient cause for that purpose. They did not, however, abandon Aristotle's external and internal causal explanations of celestial motion, but elaborated them far beyond anything that Aristotle would have recognized.

Although God, who was also the Prime (Unmoved) Mover, could move the celestial orbs directly as an efficient cause, medieval natural philosophers believed that he chose instead to delegate this task to a secondary cause of his own creation. Most assumed further that he had assigned an external, separate intelligence, or angel (the terms were usually synonymous), to move each orb, although a minority were convinced that God had opted for an internal force to keep the celestial orbs in perpetual motion.

*External movers.* How could an immaterial intelligence move a huge, cosmic orb composed of celestial ether? It was often assumed to do so by means of three things: its intellect, its will, and a third spiritual entity that carried out the command of the intellect and will, namely, a finite motive force (*virtus motiva finita*) also called an "executive power" (*potentia executiva*). This third force was required because the intellect could command the will, but the will could not carry out the act that it wills. To execute a command of the will, an executive power, or third immaterial force, was required. An intelligence, or angel, was limited in power and could not will its commands from a distance. Despite its lack of dimension and magnitude, however, an angel was thought to occupy a finite place, although it was not necessarily co-extensive with that place. That is, it could either fill the whole place or contract itself into any fraction of it. Hence the intelligence of an orb had to be in direct contact with its orb and therefore somewhere in it, or on it. Its exact location was rarely pinpointed, so that it might be anywhere within the boundaries of the orb, either at some particular point or locale or spread over the entire orb.

Because intellect and will were associated with voluntary acts, celestial motions were viewed as voluntary actions. It seemed plausible, therefore, to assume that each intelligence voluntarily commanded its orb to move with uniform, circular motion. Furthermore, because each uniform and regular celestial motion is destined to continue into the indefinite future, each motive intelligence was assumed to possess an inexhaustible force (*vir infatigabilis*), which it derived from the Prime Mover, either all at once, when he created it, or in increments, doled out as needed from his own inexhaustible reservoir.

Although intelligences were originally incorporeal, spiritual entities, they gradually assumed the status of impersonal forces. Intellect and will were de-emphasized, whereas the finite motive force, or executive power, was viewed as the real mover. Intelligences continued to play a role as celestial movers well into the seventeenth century. Even after hard orbs were rejected, intelligences were simply transferred to the planets themselves and regarded by many as their movers.

*Internal movers.* A few natural philosophers rejected intelligences as celestial movers and sought the cause of celestial motions in impersonal, internal forces. Already in the thirteenth century, John Blund and Robert Kilwardby argued that each celestial orb possessed a natural, intrinsic capability for self-motion, an opinion that might have come directly from Aristotle. By contrast with the vague, innate capacity postulated by Blund and Kilwardby, John Buridan applied his well-quantified impressed force, or impetus, theory to explain celestial motions. Because the Bible made no mention of intelligences as celestial movers, Buridan dispensed with them and assumed that at the creation God impressed incorporeal forces, or quantities of impetus, into each orb. In the absence of external resistances and contrary tendencies in the heavens, the impressed impetus of an orb would remain constant and move its orb with uniform, circular motion forever.

*Internal and external movers combined.* Even before John Buridan offered his explanation, Franciscus de Marchia (ca. 1290–d. after 1344) combined angels and impressed forces to explain celestial motions. Sometime around 1320, Franciscus assumed that an angel moved its orb by impressing a certain power (*virtus impressa*) into it. Thus, instead of the motive power operating within the angel or intelligence, Franciscus has the angel impressing a motive force into the orb with which it is associated. The impressed force then moves the orb directly. Franciscus de Marchia's solution was destined for further debate by scholastic authors in the sixteenth and the seventeenth century. The influential Coimbra Jesuits adopted it in their commentary on Aristotle's *On the Heavens* in 1592.

## Does the earth have a daily axial rotation?

Although the earth's central location in traditional cosmology was not seriously challenged until Nicholas Copernicus (1473–1543), a Polish astronomer and canon of the cathedral chapter of Frauenburg in the Kingdom of Poland, proposed his heliocentric system in the sixteenth century, its alleged condition of total rest at the center of the universe was carefully reexamined in the fourteenth century. Of the kinds of motion that could be attributed to the earth, the most important for the history of science concerned a possible daily axial rotation to explain the risings and settings of all celestial bodies.

The collective authority of Aristotle, Ptolemy, and the Bible had guaranteed unanimous acceptance of the belief that the earth lay immobile at the center of the universe (this was true despite the need to displace the earth in the eccentric systems that astronomers used). However, the possibility that the earth might rotate on its axis had been proposed and defended in Greek antiquity by Aristarchus of Samos and reported by other authors. In medieval Europe, it was a discredited opinion but was well known to all who studied and taught at the universities, or who read cursorily in astronomy and cosmology. Although it remained an unacceptable opinion in the Middle Ages, it received a surprising degree of support from John Buridan and Nicole Oresme, who discussed the problem in their questions and commentaries on Aristotle's *On the Heavens*.

As John Buridan recognized, the problem was one of relative motion. Although it appears to us that the earth on which we stand is at rest, and the Sun is carried around us on its sphere, the reverse might be true, since the observed celestial phenomena would remain the same. If the earth did actually rotate, we would be unaware of its motion. The situation would be analogous to that of a person on a moving ship that passes another ship actually at rest. If the observer on the moving ship imagines himself at rest, the ship actually at rest would appear to be in motion. Similarly, if the Sun were truly at rest and the earth rotated, we would perceive the opposite. On strictly astronomical grounds, Buridan believed that either hypothesis could save the celestial phenomena.

Buridan also added a few nonastronomical arguments, or "persuasions," as he called them. These arguments were not demonstrative, but did appear plausible. If rest is a nobler state than motion, as was often assumed, would it not be more appropriate for the nobler celestial bodies, including the sphere of the fixed stars, to remain at rest, while the earth, which was regarded as the most ignoble body in the cosmos, rotated? Because nature was usually assumed to operate in the simplest manner, would it not be simpler, and therefore more appropriate, for the small earth to turn with the swiftest speed, while the incomparably large celestial orbs remain at rest? Simplicity is again satisfied because the earth would need to rotate with a much smaller daily speed than the huge celestial orbs.

Despite these and other arguments in favor of a daily terrestrial rotation, Buridan opted for the traditional opinion. In rejecting the arguments about rest and motion, he explained that if all things were equal, it would indeed be true that

> it is easier to move a small body than a large [one]. But all things are not equal, because heavy, terrestrial bodies are unsuited for motion. It would be easier to move water than earth; and even easier to move air; and by ascending in this way, celestial bodies are, by their natures, most easily movable.[4]

Buridan's major argument against the earth's axial rotation rested on his impetus theory and certain observational consequences derived from it. He argued that the earth's rotation could not explain why an arrow shot vertically upward always falls to the same spot from which it was projected. If the earth really rotates from west to east, it ought to rotate about a league to the east while the arrow is in the air, so that the arrow should fall to the ground about a league to the west. Because we fail to detect such a consequence, it seems to follow that the earth does not rotate. But what if the earth does rotate, and the air that surrounds it rotates with it, sweeping the arrow along? Under these circumstances, the common rotatory motion shared by earth, air, arrow, and observer should result in the arrow falling to the same place from whence it was shot.

Because of his impetus theory, Buridan found this explanation unacceptable. When the arrow is projected upward, a sufficient quantity of impetus is impressed into the arrow to enable it to resist the lateral push of the air, as the latter accompanies the earth's rotation. By resisting the lateral push of the air, the arrow should lag behind the earth and air and fall noticeably to the west of the place from which it was launched. Because this is contrary to experience, Buridan concluded that the earth is at rest. For Buridan, a physical argument, not an astronomical one, decided the issue.

Nicole Oresme's discussion of the same theme is even more impressive because of his emphatic declaration that he could find no good reasons for choosing either alternative, although, in the end, he chose the traditional opinion on nonscientific grounds. His arguments in behalf of the earth's axial rotation are brilliant. In response to the argument that in ordinary experience we "see" the planets and stars rise and set and therefore infer their actual motion, Oresme, like Buridan, appeals to the relative motion of ships. Oresme reinforced the relativity argument by adding that if a man were carried round by a daily motion of the heavens and could view the earth in some detail, it would appear to him that the earth moved with a daily motion and the heavens were at rest, just as it seemed to observers on earth that the heavens moved with such a motion. Moreover, to the claim that if the earth turned from west to east, a great and easily detectable wind should constantly blow from the east, Oresme countered that the air would rotate with the earth and therefore not blow from the east.

Another experiential appeal, which Oresme attributes to Ptolemy, is similar to Buridan's arrow experience. If an arrow is shot upward, or a stone is thrown upward, as the earth rotates from west to east, would the arrow or stone fall to the west of the place from which it was launched? Because we fail to observe such effects, Ptolemy concluded that the earth is immobile, and Buridan, arguing from his impetus theory, agreed with Ptolemy. Oresme, however, saw nothing incompatible about the return of the arrow or stone to the same place from which it was launched *and* the axial rotation of the earth. To explain their compatability, it is necessary to distinguish the component motions of an arrow or stone that is hurled upward from a rotating earth. If we assume that the earth, the ambient air, and all sublunar matter rotate daily from west to east, an arrow shot into the air will have two simultaneous motions: one vertical, the other circular, the latter shared with the rotating earth. Because the arrow shares the earth's circular motion and turns with it at the same rotational speed, the arrow when shot upward will rise directly above the place from whence it was shot and then fall back to it. To an observer, who also participates in the earth's rotational motion, the arrow will appear to possess only a vertical motion. Thus the arrow would behave in exactly the same manner, whether the earth is immobile or in rotational motion. Oresme concluded that it is impossible to determine by experience whether the daily motion should be attributed to the heavens or to the earth.

Like Buridan, Oresme also presented several opinions in favor of a rotating earth that are merely persuasive, and not demonstrative. For example, a terrestrial rotation from west to east would contribute toward a more harmonious universe, since the earth and all celestial bodies would move in the same direction in periods that increase as we move outward from the earth. This would be a better solution than the traditional alternative, where two simultaneous contrary motions were assigned to the heavens, one east to west for the daily motion, the other west to east for the periodical motions. Like Buridan, Oresme also included the simplicity argument. The earth's daily rotation would be cosmically simpler because the earth's rotational speed would be much slower than the speeds required by the celestial orbs, which would have to be "far beyond belief and estimation."[5] God would seem to have made such an operation in vain.

Oresme also sought to enlist God on the side of the earth's axial rotation by reminding his readers that God had intervened on behalf of the army of Joshua (Joshua 10.12–14) by lengthening the day and commanding the Sun to stand still over Gibeon. Because the earth is like a mere point in comparison to the heavens, the same effect could have been achieved with minimum disruption by a temporary cessation of the earth's rotation, rather than halting the motions of the Sun and all other planets. Oresme suggested that in view of its greater economy of effort God might have performed the miracle in this way.

By using reason and experience, Oresme had reduced the argument to a stalemate. He was convinced that there was no more reason to choose the one alternative than the other. In the absence of demonstrative arguments for the earth's rotation, Oresme finally opted for the traditional interpretation that it is the heavens, not the earth, that rotate with a daily motion from east to west. Not only did he believe that this was in agreement with natural reason, but it also had biblical support. In the absence of demonstrative arguments for assuming the earth's rotation, Oresme would not abandon traditional arguments that were confirmed by numerous biblical passages.

Although neither Buridan nor Oresme accepted the daily rotation of the earth, some of their arguments in favor of rotation appear in Copernicus's defense of the heliocentric system, where the earth is assigned both a daily rotation and an annual motion around the Sun. Among similar arguments, we find relativity of motion, as illustrated by the movement of ships; that it is better for the earth to complete a daily rotation with a much slower velocity than would be required by the celestial orbs, which have to traverse vastly greater distances in the same time; that the air shares the daily rotation of the earth; that the motions of bodies rising and falling with respect to a rotating earth are the result of rectilinear and circular components; and, finally, on the assumption that rest is nobler than motion, that it is more appropriate for the earth to rotate than for the nobler heavens to do so.

When Oresme invoked the Joshua miracle he set a precedent for its use in subsequent discussions of the earth's rotation, especially after publication of Copernicus's *On the Revolutions of the Heavenly Spheres* in 1543. Like Oresme, both Kepler and Galileo also had occasion to explain how one might reconcile the biblical assertion that the Sun stood still to lengthen the day in favor of Joshua with the claim that the earth rotated daily on its axis. Johannes Kepler (1571–1630), one of the greatest scientists of the seventeenth century, confronted this issue in his *New Astronomy* (*Astronomia nova*) of 1609 and Galileo in his *Letter to Madame Christina of Lorraine, Grand Duchess of Tuscany* of 1615. Kepler's argument is closer to Oresme's than is Galileo's because Kepler agreed with Oresme's version that to lengthen the day and stop the apparent motion of the Sun, God stopped the earth's axial rotation. By contrast, Galileo assumed that God stopped an axially rotating Sun, which he believed was the cause of all planetary motion, and therefore brought all the planets to a halt, including the earth. But Galileo agreed with Oresme that Joshua spoke in the common language of the day, which assumed that the Sun moved around an immobile earth. Although no sustainable claims can be made that either Buridan or Oresme influenced Galileo or Kepler in any of the particulars mentioned here, we must at least credit these fourteenth-century natural philosophers with anticipating some rather interesting and important arguments.

## THE WORLD AS A WHOLE, AND WHAT MAY LIE BEYOND

Aristotle argued vigorously against those who believed in some kind of existence beyond our finite, spherical world. With the kind of physics and cosmology he had developed, extracosmic existence seemed unintelligible. In his judgment, there was nothing beyond our world: neither matter, nor time, nor void, nor place. He regarded the existence of such entities beyond our world to be naturally impossible. In an extraordinary series of departures, however, medieval natural philosophers transformed Aristotle's natural impossibilities into divine possibilities. They focused attention on three major questions about the world as a whole: was it eternal or created; were there other worlds beyond it; and did some kind of place or space exist beyond its finite boundaries?

### Is the world created or eternal?

Among the problems confronting Christian natural philosophers in the late Middle Ages, none was more difficult than that of creation. This was apparent in chapter 5, where we saw the role played by the eternity question in the Condemnation of 1277 and in the controversies that raged between theologians and natural philosophers, and among the theologians themselves. Here we will focus on the arguments.

The broad alternatives are straightforward: is the world eternal, without beginning or end, as Aristotle and some Greeks argued; or, did the world have a beginning, and would it have an end? During the early Middle Ages, support for a creation came from some pagan sources, from Plato's *Timaeus* and Macrobius's *Commentary on the Dream of Scipio*. The doctrine of creation went largely unchallenged. With the introduction of Aristotle's works in the twelfth and thirteenth centuries, powerful arguments for the eternity of the world became available. Fearful that it might undermine belief in a supernatural creation from nothing, the bishop of Paris condemned the idea of an eternal world in 1270, and then, as if to underscore its importance, condemned it again in some twenty-seven different propositions in the Condemnation of 1277. During the thirteenth century, and all through the Middle Ages, the faithful had to accept the creation of the world. Within this context, three opinions found varying degrees of support: (1) some, like Saint Bonaventure, insisted that the creation of the world was capable of rational proof; (2) others countered that the eternity of the world was rationally demonstrable; and (3) in the middle was Thomas Aquinas, who argued that no rational proof was possible for either side and, further, suggested that the world might be viewed as both created and eternal.

Bonaventure, whose proofs were derived from Islamic sources and ultimately from John Philoponus,[6] proposed a typical argument against a beginningless world and in favor of a creation. If the world had no beginning, it follows that an infinite number of celestial revolutions must have occurred to the present. Because an infinite number of revolutions cannot be completed or traversed, the present revolution could not have been reached, which is absurd. As further evidence of the absurdity of an eternal, beginningless, world, Bonaventure asserted, as a second argument, that if an actual infinite number of revolutions has occurred to the present, then all subsequent revolutions must be added to an already actually infinite number of revolutions. Adding to an infinite, however, cannot make it larger, because, as Bonaventure put it, "nothing is more than an infinite."[7] As a third type of argument, Bonaventure traded on apparent differences between infinites. If the world has an infinite past, then the Moon will have made twelve times as many revolutions as the Sun. Therefore the Moon's infinite will be larger than the Sun's, which is impossible.

Those who reacted against Bonaventure's arguments did not claim an eternal world, or deny the creation. They simply attempted to show that arguments for a beginningless world were intelligible. The Aristotelian concept of a potential infinite played a significant role. Against Bonaventure's first argument, they showed that an actual infinite number of days need not have passed. In a beginningless universe, there was no first day from which to begin the count of days. There is no actual infinite number of days, but only a potential infinite. That is, we can assume that a potentially infinite number of days have passed to the present day and that more days can be added indefinitely. Or we can imagine that an infinite sequence of days extends backward in time, because there can be no first day to halt the sequence. Thus where Bonaventure based his arguments on the concept of an actual infinite, his opponents defended their position by appeal to a potential infinite.

To counter Bonaventure's argument about the absurdity of different infinites, some larger than others, scholastic natural philosophers in the fourteenth century offered two arguments. In one, they showed that just because the Moon makes twelve revolutions to one for the Sun, does not signify that, in a beginningless world, the Moon will have traversed an infinite twelve times the size of the Sun's. John Buridan and others argued that one infinite is not greater than another. An infinity of years is no greater than an infinity of days. Every infinite is equal to every other infinite. In support of this point, some added that there are no more parts in the whole world than in a millet seed: both are infinitely divisible. The second argument against Bonaventure showed that infinites might indeed differ in size, if one was a subset of the other, as is the case of the Moon's infinity of revolutions with respect to that of the Sun.

Most scholastic natural philosophers, however, were agreed that neither the creation of the world nor the eternity of the world was demonstrable. Indeed each was equally probable. "That the world had a beginning," Thomas Aquinas insisted, "is an object of faith ... not of demonstration or science."[8] With two equally probable alternatives available, but with faith demanding acquiescence in a world that had a beginning by means of a divine creation, many – including Thomas – chose a compromise solution: the world was both created and without a beginning. The creation of the world was assumed on the basis of faith. But how could it also be eternal, that is, without a beginning? Perhaps God had chosen to will the existence of creatures without a temporal beginning. If God had indeed acted in this manner, then he must have decided to will the co-eternality of things with himself. Thomas Aquinas showed that this was possible when he declared that "the statement that something was made by God and nevertheless was never without existence, does not involve any logical contradiction."[9] Because of his absolute power, God, as efficient cause of the world, "need not precede His effect in duration if that is what He Himself should wish."[10] God could have created an eternally existing world because he produces effects instantaneously. But if God and the world are co-eternal, does this also mean that they are co-equal? That unacceptable consequence is avoided because our mutable world is assumed to be wholly dependent on the immutable deity. After all, God possesses the power to cause the material world to disappear, but the world is powerless to affect God. It is totally dependent on him.

The possibility that the world could have existed from eternity and also have been created was a surprisingly popular idea during the Middle Ages and Renaissance. We may view it as an attempt to save a foundational principle of Aristotle's natural philosophy, even though that principle could never be more than a possibility for Christian natural philosophers.

### On the possible existence of other worlds

When Aristotle discussed the possibility of other worlds, he assumed that those worlds existed simultaneously with our own and were identical with it. But he rejected the existence of other worlds, largely because of his own arguments that showed the impossibility that any thing – matter, place, vacuum, or time – could exist beyond our world. The debate in the Middle Ages was also largely about identical worlds, that is, about worlds that possessed identical elements, compounds, and species as are found in our world.

Aristotle's response to the problem of extracosmic existence proved unsatisfactory and drew criticism in the ancient world. In his sixth-century *Commentary on Aristotle's On the Heavens*, Simplicius, one of the most important Greek commentators of late antiquity, reported the reaction of Stoic philosophers to Aristotle's denials of existence beyond the world. The Stoics imagined that someone located at the extremity of the world extended an arm beyond that extremity. What would happen? They could conceive of only two alternatives: either the arm meets some obstacle and cannot be extended further, or it does not and therefore can be extended beyond the world. If the first alternative occurs, they further imagine that the individual climbs out onto the obstacle that obstructs the arm and again extends an arm. Once again, the person either meets another obstacle, or extends an arm into an empty space that lies beyond the world. The implication of Simplicius's account is clear. Matter may indeed exist beyond the world, but because our world is finite, matter cannot be extended indefinitely. Eventually, one must reach void space. Simplicius's commentary on *On the Heavens* was translated into Latin in the thirteenth century and became widely known. His brief discussion provided a historic link with an important anti-Aristotelian tradition that gave medieval natural philosophers reason to believe that it was possible that there was either matter or void space, and perhaps both, beyond the world.

The possibility that matter might exist beyond our world in the form of other worlds became a major subject of discussion as a result of the Condemnation of 1277. Prior to 1277, the possibility of other worlds was not seriously entertained by Christian authors. After all, Aristotle and his Christian followers agreed that there is only one world. One interesting question did, however, arise: although God created a single, unique world, could he have created other worlds, and could he now create other worlds, if he wished to do so? The Condemnation of 1277 dramatically altered the intellectual context at the University of Paris, and after 1277, the question about other worlds became commonplace. Article 34 was condemned because it denied that God could make more than one world. Thereafter, all had to concede that God could create as many other worlds as he pleased.

Although three kinds of cosmic pluralities were distinguished, only one was given extensive treatment, namely, the possibility that identical, simultaneous, distinct, and separate worlds existed in a void space (the other two involved successive, individual worlds and a plurality of simultaneously existing concentric worlds). Upon assuming that God had indeed created other identical worlds, natural philosophers had to confront a significant problem, one raised by Aristotle himself. Because these worlds were assumed to exist simultaneously, would the elements of one world tend to move toward the center and circumference of another world, rather than to the center and circumference of their own world? More specifically, would the earth of one world, or any part of the earth, seek the center of another world? For example, would a heavy particle of earth in another world seek the center of our world by first rising upward in its own world, contrary to its own natural tendency to move only toward its own center, and then, having reached our world, after somehow traversing the intermediate space between the two worlds, fall toward its center? If this were possible, heavy bodies would appear to possess two contrary natural motions, which Aristotle thought absurd. Similar reasoning applies to fire, which might rise in our world and, upon reaching another world, descend to its natural place between air and the lunar sphere. Fire would therefore be capable of rising and falling naturally. The same heavy or light body would be capable of contrary natural motions, which violated two of Aristotle's principles: that a simple heavy or light body could have only one natural motion, and that only one center and circumference of the world could exist. Thus could a purely hypothetical, counterfactual question pose meaningful problems for natural philosophers.

In rejecting Aristotle's opinion, Richard of Middleton expressed a widely held interpretation. Even with identical worlds, Richard argued, neither the earth nor any of its parts would rise in its own world in order to reach the center of another world. To the contrary, the earth of each world would remain at rest in the center of its own world. Any parts that were removed from the center would, if unimpeded, always return to that same center. Each world was viewed as a self-contained system with its own center and circumference, uninfluenced by any other world. Thus if it were possible to remove the earth of another world and place it at the center of ours, that earth would thereafter remain at rest in the center of our world. Conversely, if earth from our world were removed to the center of another world, it would remain at rest in its new center and have no inclination to return to its former place. Following the Condemnation of 1277, many deemed what was impossible in Aristotle's natural philosophy to be possible and plausible. Aristotle's physics and cosmology could operate in numerous worlds, if God chose to create those worlds. To accept this possibility, Aristotle's idea that only one center and circumference could exist was abandoned in favor of a possible multiplicity of centers and circumferences – one pair for each world. Moreover, all centers would be equal and none unique, thus also casting doubt on Aristotle's doctrine of natural place for the four elements, which depended on a unique world with a single center and circumference.

During the Middle Ages, there were no actual supporters of the existence of other worlds. It was apparently sufficient to demonstrate that if God made other worlds – and it was always conceded that he could do so – they would be subject to the same physical laws as our world. Thus where Aristotle regarded the existence of other worlds as absurd and impossible, his medieval followers thought such worlds were possible and intelligible – even if only by divine command.

### *Does space or void exist beyond our world?*

The possibility that God could create other worlds, coupled with Aristotle's definition of a vacuum as a place deprived of body but capable of receiving it, implied that void space might exist beyond our world. If other worlds were created, it seemed plausible to assume that void space intervened between them. In discussing the two extracosmic entities – worlds and void space – an important distinction must be made. Other worlds were mere possibilities that gave rise to significant discussions about the Aristotelian world. With extracosmic void space, however, real existence was sometimes assumed, occasionally in terms of natural arguments, but more often because of supernatural reasons. The significant developments about void space in the Middle Ages were associated with God and the supernatural. The mere possibility that God could create another world was used by Robert Holkot (d. ca. 1349), a Dominican friar, to argue for the existence of something real beyond our world. He first inquired about whether something exists beyond our world. If it does, then we can declare that something does indeed exist beyond our world. But if nothing exists beyond our world, and yet some body could possibly exist there – for it was assumed that God could create another world there – it follows that "beyond [our] world there is a vacuum because where a body can exist but does not there we find a vacuum. Therefore a vacuum is [there] now."[11] Thus the mere possibility that God could create another body, or world, beyond our world – even if he did not actually create it – implies the existence of an actual vacuum. Holkot furnished a vivid illustration of how a counterfactual could have real consequences for a natural philosopher.

It was not the arts masters who would accept the existence of extracosmic void, but the theologians. Discussions about extracosmic void space fell naturally to the theologians because ideas about infinite space grew naturally from Christian concerns about God's location in the world, and his immutability. Among medieval theologians who discoursed about an extracosmic vacuum, none was more important than Thomas Bradwardine, who assumed the existence of an infinite void space and derived its properties from the attributes of God. In his treatise *In Defense of God against the Pelagians (De causa Dei contra Pelagium)*, Bradwardine argued for God's ubiquity. But he then inquired whether that ubiquity is confined to the world, as Saint Augustine and others believed, or extended beyond. It cannot be confined to the world, because prior to its creation God must have existed eternally in the place in which the world would be created. If not, then God would have had to arrive there from elsewhere, which is impossible, because God is immutable and does not move from one place to another. To attribute motion to God would reduce his status from perfect to imperfect. Because God could have created the world in any void space he pleased, and because he does not move from one place to another, we must assume the existence of an infinite number of void spaces in any one of which God might have created the world and in all of which he has existed eternally. Taken collectively, these places constitute an infinite, imaginary void space in which God is omnipresent. In support of this conclusion, Bradwardine reasoned that God would be more complete and perfect if he existed in many places simultaneously rather than in a unique place only, as would be the case if he confined himself solely to the world he created.

How did Bradwardine envision the relationship between an eternally existing infinite, void space and the God who occupied it? Was infinite space independent of God? Were God and infinite space two co-eternal, co-existing, independent entities? Such a relationship was unacceptable. It would compromise and diminish God's unique status. The solution was to make infinite void space and God one and inseparable by assuming that infinite void space was God's infinite immensity. The relationship was God's way of establishing his infinite omnipresence.

How could God be "spread out" in an infinite space without being somehow extended? Does God's omnipresence in an infinite void space imply that he is an extended, three-dimensional being? Such a thought was repugnant to medieval theologians. Bradwardine therefore insisted that God "is infinitely extended without extension and dimension."[12] But if infinite, void space is identified with God's immensity, which "is infinitely extended without extension and dimension," it appears that Bradwardine intended to deny extension and dimensionality to infinite void space.

In arguing that God was omnipresent in an infinite void space, Bradwardine opposed the views of Saint Augustine and Thomas Aquinas, whose opinions agreed with Aristotle's. Where Bradwardine argued that God can act anywhere because he is everywhere, John Duns Scotus and his followers insisted that God's will, not his omnipresence, was the basis of divine action. God need not be in a place to act on it. He only needs to will an action for it to occur, and he can exercise that will on, and in, a place remote from his actual presence. God's presence in the place where he created the world is unnecessary because he could have willed the world's creation from some other place. Thus did Duns Scotus deny the necessity of God's presence in an infinite void space.

That Bradwardine's void space was unlike any other earlier description of a vacuum is apparent when he proclaims that "void can exist without body, but in no manner can it exist without God."[13] How different from the infinite vacuum envisioned by Greek atomists and Stoics, who assumed a three-dimensional space devoid of both body and spirit. The cosmic configuration adopted by the Stoics did, however, bear a striking resemblance to Bradwardine's vision: a spherical, finite world that is a plenum and that is surrounded by an infinite void space. Those scholastic natural philosophers who assumed an extracosmic infinite void space also adopted the same conception of the universe.

Although Bradwardine's ideas were all but forgotten until the publication of his work in 1618, he was apparently the first to link God and infinite void space and then to consider that space as real. As God's immensity, infinite void space had to be real. But if God was spread out in an infinite space, did this not imply that he was somehow divisible and composed of parts? For example, was the part of God lying in some part of the void beyond the world identical with that part of God associated with a planet, or an orb? It was difficult to speak of God's omnipresence without invoking the language of magnitude and quantity. The solution to this vexing problem did not come from Bradwardine but had already been formulated in the thirteenth century (around 1235) by Richard Fishacre, in his theological commentary on the *Sentences* of Peter Lombard. Richard argued that God's infinite immensity always remains indivisible because he is wholly and indivisibly in every part of space, an interpretation that came to be called the "whole-in-every-part" doctrine. Because God was wholly in every part of space, however small or large, he was not divisible. Whether the space that was perceived as God's immensity was also indivisible is unclear, although bodies occupying parts of that infinite space within our world were indeed divisible.

Development of the concept of an omnipresent God in an infinite void beyond the created, vacuumless world cannot be understood in isolation from the powerful intellectual currents unleashed by the Condemnation of 1277. The emphasis on God's absolute power influenced fourteenth-century thought in theology, philosophy, and natural philosophy. In the intellectual climate generated by the condemnation, it would have seemed strange to suppose that the presence of an all-powerful God should extend no further than the finite cosmos that he created, simply because Aristotle had denied extramundane existence on the basis of the inner logic of his system. His arguments against it involved the kind of necessities and impossibilities that had fallen under a dark cloud in the theological and philosophical thought of the fourteenth century. The emphasis on hypothetical arguments often inspired new looks at possibilities that were in conflict with Aristotelian physics and cosmology.

A good example of this with respect to void space beyond the world is article 49 of the Condemnation of 1277, which declared that "God could not move the heavens [or world] with a rectilinear motion; and the reason is that a vacuum would remain." It was now necessary to concede that God could indeed move the heavens, or the world, with a rectilinear motion. If God chose to do so, however, three major Aristotelian principles would be simultaneously violated: (1) the rectilinear motion of the world would leave behind a vacuum, which is impossible in

Aristotle's world; (2) the motion would not be classifiable under any of the three natural motions that Aristotle had distinguished, namely rectilinear up or down motions, or circular motion; and (3) the rectilinear motion would be independent of places, because Aristotle had argued that in the absence of matter beyond the world, no places can exist there.

If the motion of the world leaves a vacuum behind, an obvious implication is that the world is located in a vacuum, an unpleasant consequence for Aristotelians. Although a rectilinear motion of the whole world made no sense in the Aristotelian system, it raised important questions about a motion that could occur in the absence of places. Nicole Oresme believed that if God chose to move the world rectilinearly, it would be an instance of an absolute motion, one that was independent of places. Under the conditions of article 49, the spherical cosmos would be the only body in the universe. Because the body in motion could not be related to any other place or body, its motion would be absolute in a real but nondimensional, infinite, void space. In 1715, Samuel Clarke insisted, against Gottfried Leibniz (1646–1716), that if God moved the entire world with a rectilinear motion, that motion must be considered a real motion in an absolute, real space, a position that was virtually the same as Oresme's, except that Clarke's space was three-dimensional.

Medieval ideas about infinite void space had an impact not only on scholastic authors of the sixteenth and seventeenth centuries but also on major nonscholastic authors, such as Otto von Guericke (1602–1686), Henry More (1614–1687), Samuel Clarke (1675–1729), Isaac Newton (1642–1727), and Joseph Raphson (d. 1715 or 1716). The split between the medieval and the early modern scholastic interpretations of infinite space, and that of the nonscholastic authors just mentioned, concerned the nature of space and the God that filled it: were they dimensional or nondimensional? Ideas about the vacuum drawn from the ancient world, from experiments on atmospheric pressure, and from the construction of artificial vacua, led nonscholastic scientists and philosophers inexorably to think of void space as three-dimensional. Many of them then had to judge the nature of the God that was omnipresent in that three-dimensional void space. Some inferred that he was as three-dimensional as the space he occupied. Henry More, Isaac Newton, Joseph Raphson, Samuel Clarke, and Benedict Spinoza (1632–1677) were among those who concluded that, in order to fill an infinite, three-dimensional, void space, God himself had to be a three-dimensional extended being. For example, Joseph Raphson, a mathematician and Fellow of the Royal Society, believed that only if God was truly extended in space could he be omnipresent. Raphson spoke disapprovingly of scholastic natural philosophers and theologians who assigned a transcendent sense of extension to God. But how, Raphson argued, can extended beings come from something that is only transcendentally, and not actually, extended? For

Raphson, and numerous others that he names, God is an infinitely extended being whose immensity is infinite void space. Although God was characterized as a three-dimensional, infinite being, More, Newton, Raphson, and others thought of him as immaterial. It was Benedict Spinoza who took the final step and converted the deity into a three-dimensional, infinite, material, and corporeal entity. The divinization of space, which began in the late Middle Ages, was initially nondimensional, or transcendent, as Raphson would express it. By the time this infinite divinized void space became the space of Newtonian physics, the God who occupied it, and whose attribute it was, had been transformed into a corporeal being.

Based on the examples cited in this chapter, we may rightly conclude that the departures from Aristotle's ideas and principles were numerous and important. The new spatial concepts alone represent dramatic departures. Rather than significantly altering Aristotelian natural philosophy, however, these changes became part of it. Without critical analysis, Aristotelians absorbed these ideas into the larger whole of Aristotelianism. But what was this larger whole of Aristotelianism? Now that I have characterized Aristotle's natural philosophy (chapter 4) and its reception in the West (chapter 5), and given a sense of how medieval natural philosophers altered their Aristotelian legacy (this chapter), I shall, in chapter 7, describe the essential features of medieval natural philosophy, and of the Aristotelians who shaped it into the phenomenon we call Aristotelianism.

# 7

# Medieval natural philosophy, Aristotelians, and Aristotelianism

BECAUSE Aristotelian natural philosophy is the major emphasis in this volume, our discussion of it must necessarily encompass Aristotle's natural books (*libri naturales*), as described in chapter 3, and the medieval commentaries and questions on those works. The natural books of Aristotle were far from a thorough, well-rounded, coherent, and systematic description and analysis of the physical world. But in those treatises, a wealth of topics and ideas were included, and a remarkable breadth of coverage. The natural books were the best available guides for the study of the universe, which is why they served as the fundamental texts for natural philosophy in the universities of the Middle Ages. It was that natural philosophy that functioned as the world view of the Middle Ages, a world view that was embodied in a special kind of literature – the questions literature – that was peculiar to the Latin Middle Ages and to the medieval university.

## THE QUESTIONS LITERATURE OF THE LATE MIDDLE AGES

The *questio*, or question, was the most widely and regularly used format for natural philosophy. As we saw in chapter 3, it grew out of the commentary, but it was structurally akin to the oral disputation that was such a prominent feature of medieval university education. It was actually a teaching master's written version of the questions that he presented orally in his classroom lectures. Because of its disputational structure, the *questio* form of literature and analysis has become almost synonymous with the concept of medieval "scholastic method." Although occasional variations in the arrangement of the constituents of a *questio* occurred, scholastics tended to present their arguments in a rather standard format that remained remarkably constant over the centuries. First came the enunciation of the problem or question, usually beginning with a phrase such as "Let us inquire whether" or simply "Whether" (*utrum*): for example, "Whether the earth is spherical," or "Whether the earth moves," or "Whether it is possible that several worlds exist." This was followed by one or more – sometimes as many as five or six – solutions supporting either the negative or the affirmative position. If the arguments for the affirmative side appeared first, the reader could assume that the author would probably adopt the negative position; conversely, if the negative side appeared first, it was likely that the author would subsequently adopt and defend the affirmative side. The initial opinions, which would ultimately be rejected, were called the "principal arguments" (*rationes principales*).

Immediately following the principal arguments, the author presented, briefly, the opposite opinion, usually introduced by the term *oppositum*. The appropriateness of using the term "opposite" to introduce the alternative opinion is evident from the fact that medieval authors were responding to questions that required yes or no responses. Thus, if the opening principal arguments of a question represented an affirmative, or yes response, then the following opposite response must represent the negative, or no position. The *oppositum* was largely confined to the citation of at least one authority – often Aristotle himself – who was in disagreement with the opening affirmative opinion. Indeed the opposite opinion might initially be encapsulated in a statement as brief as "Aristotle determines the opposite."

After the principal and opposite arguments, the author might further clarify and qualify his understanding of the question or explain particular terms in it. For example, in a question asking "Whether it is possible that several worlds exist," John Buridan explained not only that the term "world" could be taken in many ways but also what he meant by "several [i.e., a plurality of] worlds":

> "world" (*mundus*) can be taken in many ways. In one way as the totality of all beings; thus the world is called "universe" (*universum*). "World" is taken in another way for generable and corruptible things and in another way for perpetual things; and so it is that we distinguish world into this inferior world and into a superior world. And yet "world" is taken in many other ways that are not relevant to our present discussion. But "world" is taken in another way that is pertinent, [namely,] as the totality of heavy and light [bodies]. And it is about such a world that the question – whether it is possible that several worlds exist – inquires. And with regard to this, it must be noted that a plurality of such worlds can be imagined in two ways: in one way existing simultaneously, as if outside this world one other such world existed now; in another way they exist successively, namely one after the other.[1]

After adding whatever qualifications he deemed necessary, the author was ready to present his own opinions, usually by one or more detailed conclusions or propositions. To anticipate objections, the master might even choose to raise doubts about his own conclusions and subsequently resolve them. To conclude the question, he would respond sequentially to each of the principal arguments enunciated at the outset.

A typical questions treatise consisted of a considerable number of questions, each structured in approximately the manner just described. In the fourteenth century, Albert of Saxony included 107 questions in his *Questions on the Eight Books of Aristotle's Physics* and 35 in his *Questions on the Two Books of On Generation and Corruption;* John Buridan considered 59 questions in his *Questions on the Four Books of Aristotle's On the Heavens* and 42 in his *Questions on the Three Books of Aristotle's On the Soul;* and Themon Judaeus treated 65 questions in his *Questions on the Four Books of Aristotle's Meteorology*. In these five distinct treatises, three authors considered 308 different questions.

Each of Aristotle's treatises was thus divided into a series of questions. Generally, the questions on a given Aristotelian treatise followed the subject matter of that treatise. Because Aristotle's treatises were themselves loosely organized and often unintegrated, the questions that evolved from those treatises during the late Middle Ages were similarly lacking in cohesion. Although, as we shall see, sporadic efforts were made to link questions, it was more usual to treat each question independently, as if it were unrelated to any other question or topic. In medieval, scholastic, natural philosophy, the focal point was on the independent question and the differing opinions that it generated. The object was to resolve, or determine, each question.

In the numerous questions that comprised a typical questions treatise, scholastic authors made frequent references to Aristotle's arguments and opinions, whether those references were to various parts of the work on which they were writing questions, or to another of Aristotle's works. It was not customary, however, to relate substantive matters in one question to similar discussions in any other question, either in the same treatise or in some other treatise. Nevertheless, such connections were made in two basic ways. Sometimes authors of scholastic questions made references to other questions in the same treatise. For example, they might say: "as is sufficiently obvious from another question," or "the opposite was said in another question," or "as was seen in another question," and so on. The difficulty in locating such vague references is obvious. More common, however, was an even vaguer kind of referral, where an author might allude to relevant thoughts that occur earlier and later in the treatise. In this category are expressions such as "as was said earlier," or "as was touched upon," or "as will be seen later," or similar variants.

Although most scholastic natural philosophers were sparing, and perhaps even haphazard, in the linkages they indicated, some were diligent in their efforts to connect with arguments made earlier and later. Eminent authors such as John Buridan and Albert of Saxony made relatively few cross-references. Nicole Oresme, however, was a notable exception. In his lengthy *Questions on the Books of Aristotle's On the Soul (De anima)*, Oresme presented forty-five questions spread over the three books. In these forty-five questions, Oresme made reference to other questions approximately twenty-five times (in the form of "in another question," or "as was seen in another question," or "as is obvious in the preceding question," and so on). Oresme also found approximately seventy occasions to use the vaguer kind of reference mentioned earlier, as when he declared: "as was said," or "as was seen," or "in the same way as was previously stated," or "as was declared elsewhere," or "as will be seen later," and so on.

What do these cross-references reveal about Oresme and his treatise? At the very least, that Oresme was an author who sought to inform his readers of relevant material in other questions elsewhere in the treatise. Does this suggest that Oresme sought seriously to integrate his *Questions on Aristotle's On the Soul* into a cohesive whole? Examination of his references fails to reveal such a larger objective. Not only are most of the approximately ninety-five references difficult to determine, but it is not always certain that they are to statements in the same treatise, or that they occur at all. The editor of Oresme's text, Peter Marshall, observed in one place, that no subsequent discussion follows Oresme's pronouncement that "this will be discussed elsewhere." On another occasion, Oresme used the phrase "as was touched on previously," and the editor cited four separate questions in which Oresme seems to have "touched on" that subject.

Perhaps the most obvious difficulty with Oresme's references is their vagueness. How many readers would be sufficiently dedicated or energetic to search for references that proclaim "as was said in another question" or "as was said above"? But even if a determined reader located a given reference, what was to be made of it? Oresme rarely explained the connection between a given passage and related texts. His readers were left on their own, first to find relevant passages that might illuminate a given text and then to determine their relationship.

Why did Oresme and his scholastic colleagues give such vague references? After all, the questions are numbered in the text. For example, in the third book of his *Questions on Aristotle's Book On the Soul*, Oresme might have made reference to one or more numbered questions. And yet, in all of his ninety-five cross-references, not a single one is to a numbered question. Nowhere does Oresme refer, for example, to the fourth question of the third book, or the eighth question of book two, and so on. It seems extraordinary that he and his colleagues would have missed the most obvious and helpful way to make cross-references. But even if such specific references had been found, they would have been of little value without some effort to connect the passages involved, some effort to explain their relationship. But neither Oresme nor his scholastic colleagues thought of integrating their questions in this way.

Primacy in a medieval questions treatise was accorded to the individual question, which was to be treated thoroughly and then left as a definitive statement of the author's position. The tendency was to treat each question in isolation from other questions, even though there might be numerous linkages between questions in a given treatise, and also to questions in other treatises. Even when these were indicated, as Oresme did some ninety-five times in a single questions treatise, they were only marginally useful because the references were not only vague and uncertain, but their alleged connections were left unspecified.

Authors of questions treatises sought to analyze each question into its constituent parts, and not to synthesize the questions into a larger whole. Medieval natural philosophers laid emphasis on analysis, not synthesis, and preferred to follow Aristotle's order of topics, rather than organize the questions into some more meaningful, larger picture, of the world. Although the questions format was useful for coping with narrow, specific problems, it was inadequate for broader, interrelated themes, or for any subject that required a sustained presentation. In his *Questions on Aristotle's Book On the Soul*, Nicole Oresme complained (book 3, question 3) that in treating the various aspects of the human intellect, the questions format was inadequate, but added that he would proceed nonetheless because custom required it. Among scholastic authors, Oresme is unusual for the number of tractates (*tractatus*), or treatises, he wrote on a variety of subjects, in each of which he could treat a given subject in systematic detail, as he did in his treatises on proportionality, on intension and remission of forms, on the commensurability or incommensurability of celestial motions, and on the sphere. He would have regarded these themes as unsuitable for the questions format.

### NATURAL PHILOSOPHY IN OTHER LITERARY MODES

Questions in natural philosophy were not only presented in treatises that were actually called "questions" on a particular Aristotelian treatise, but they also appeared in commentaries on the *Sentences* of Peter Lombard, especially in the second book, which treated the creation and usually included questions on celestial structure and operations, as well as on light. Because they were also concerned with creation, "summas" in theology included questions similar to those found in commentaries on the *Sentences*.

Summas in theology and *Sentences* commentaries were logically arranged genres of scholastic literature (Thomas Aquinas's *Summa theologiae* is a supreme example). One might expect that a summa in natural philosophy would be similarly systematized. During the first quarter of the fifteenth century, Paul of Venice (ca. 1370–1429) composed *Summa of Natural Philosophy* in which he divided natural philosophy into six parts, corresponding, respectively, to Aristotle's *Physics, On the Heavens, On Generation and Corruption, Meteorology, On the Soul,* and *Metaphysics.* The order in which he arranged the first four treatises was undoubtedly based on Aristotle's opening remarks in the *Meteorology.* As a faithful Aristotelian, Paul followed Aristotle's order of works. And as a traditional scholastic natural philosopher, he also treated the questions in each treatise in isolation from the questions in the other treatises. Paul's *Summa of Natural Philosophy* is therefore little more than a collection of six distinct Aristotelian treatises, each with a set of its own questions. Paul of Venice achieved no more integration and synthesis than if he had published the questions on each treatise separately.

The closest medieval scholastics came to a systematic synthesis of the "big picture" occurred in treatises that did not include questions. These treatises, known under the generic name of "compendia," were attempts to explain Aristotle's opinions about the whole range of the world's operations, and to do so in a coherent, logical, and reasonably brief manner. One of the best exemplars of this genre is an anonymous treatise on natural philosophy and metaphysics that was probably composed in the second half of the fourteenth century at Paris.[2] At the outset, the author explained that because Aristotle's texts were so prolix and difficult for students to read he had seen fit to summarize Aristotle's opinions, and those of other philosophers, within the compass of a brief compendium (although with a total of 236 folios, it can hardly be considered brief). The treatise takes on added significance because the anonymous author said that many scholars had written on topics that Aristotle had barely discussed, or had not mentioned at all. Thus in his summary, this author included not only Aristotle's interpretation but also departures from Aristotle by "moderns," that is, by authors who probably wrote in the thirteenth and fourteenth centuries. This unusual treatise covers a wide range of natural philosophy. Its author sought to organize the topics systematically and to introduce new opinions and departures from Aristotle where these seemed relevant. The result was a text that was much more informative than the standard questions treatise. Such compendia, however, were relatively rare in the Middle Ages and, by comparison to treatises in the questions genre, played a minor role.

There was, however, much natural philosophy embedded in non-questions works that included the word "treatise," or "tractate" (*tractatus*), in their titles. One of the most important was John of Sacrobosco's *Treatise on the Sphere,* a thirteenth-century work that briefly described the heavens and the earth and was used as a university textbook. It gave rise to a number of commentaries and independent treatises titled *On the Sphere.* As mentioned earlier, Nicole Oresme wrote some of the most significant works on natural philosophy using the treatise format, as we find in his *Treatise on the Configuration of Qualities and Motions (Tractatus de configurationibus qualitatum et motuum), Treatise on Ratios of Ratios (Tractatus de proportionibus proportionum)*, and *Treatise on the Commensurability or Incommensurability of the Celestial Motions (Tractatus de commensurabilitate vel incommensurabilitate motuum coeli)*. Encyclopedic works by three thirteenth-century authors, William of Auvergne (ca. 1180–1249) (*On the Universe [De universo]*), Bartholomew the Englishman (fl. 1220–1250) (*On the Properties of Things [De rerum proprietatibus]*), and Vincent of Beauvais (ca. 1190–ca.1264) (*The Mirror of Nature [Speculum naturale]*), contained much about natural philosophy and were influential throughout the Middle Ages. As mentioned in chapter 3, natural philosophy was often a significant component of works on medicine, theology, moral philosophy, and metaphysics. Indeed, it also formed an integral part of alchemical treatises, where the nature of the elements was considered within the broader framework of matter theory. Although the focus in this study is on the questions literature, because it was the most important source for medieval natural philosophy and the world view it depicted, other literary forms, especially the independent tractates, also played a role.

## THE COSMOS AS SUBJECT MATTER OF NATURAL PHILOSOPHY

The purpose of the literature of natural philosophy was to describe and analyze the structure and operation of the cosmos, with all its objects and creatures. In chapter 6, I focused on the departures medieval natural philosophers made from Aristotle's world view, as described in chapter 4. In the next few paragraphs, I shall briefly describe a reasonably representative medieval world view. As with most "world views," the medieval version had two fundamental but interrelated aspects. The first, often equated with the medieval world view to the exclusion of the second, involved the overall structural frame of the cosmos – the macro structure, or the "big picture." The second aspect focused on the details of cosmic operations, where we find the largest degree of controversy and disagreement.

### The big picture

The cosmic frame was, on the whole, remarkably simple. It was a composite, made up largely of cosmological materials drawn from Aristotle's natural philosophy but also including ideas from scriptural texts, especially the creation account in Genesis, as well as from traditional concepts and dogmas about the deity, angels, and souls, that had evolved within

Christian theology. The cosmos was an enormous, unique, finite, material sphere filled everywhere with matter. The sphere itself was divided into numerous subspheres, or orbs, nested one within another. Within this huge sphere and its subspheres, existed two radically different parts: the celestial and the terrestrial. The former began with the concave surface of the lunar sphere and ascended all the way to the sphere of the fixed stars, and even beyond to the empyrean heaven, the outermost sphere of the world, where the blessed were assumed to live in luminous splendor. The celestial region was filled with a perfect, incorruptible ether, which had as one of its primary properties the ability to move itself with uniform circular motion, or had the capacity to be moved by something else, say an intelligence or angel. Because they were composed of this remarkable ether, the concentrically arranged celestial orbs – usually assumed to be anywhere from eight to eleven in number – moved around the center of our spherical universe with uniform circular motions, carrying the fixed stars and the seven planets. Eight of the orbs carried celestial bodies: the eighth bore all the fixed stars, and the seven below it carried the seven planets, one to a sphere.

The terrestrial region, which began just below the concave surface of the lunar sphere, descended to the geometric center of the universe. In contrast to the celestial region, the terrestrial, or sublunar realm, was characterized by incessant change, as the imperfect and corruptible bodies within it continually came into being and passed away. These terrestrial bodies were compounded of four elements that were arranged in a series of four concentric orbs, each of which served as the natural place of one element. In descending order from the concave lunar surface, the first orb was the natural place of fire; the second of air; the third of water; and the fourth of earth. Each element possessed an innate capacity for natural motion toward its own natural place. The dominant element in any body determined the direction of the body's natural motion, which was always toward the natural place of the dominant element. When unimpeded, earthy bodies, heavy by nature, always fell naturally toward the center of the universe, whereas fiery bodies, which were regarded as absolutely light, rose toward the lunar concavity. The intermediate elements, water and air, produced dual effects, depending on their locations: watery bodies rose in the natural place of earth and fell in the natural place of fire, whereas airy bodies rose in the natural places of earth and water and fell when located in the region of fire.

Because the celestial region was held to be incorruptible, it was judged more perfect, and therefore nobler, than the terrestrial region. On the basis of an almost unanimously accepted principle that a nobler body can affect and influence a less noble body but that the reverse is not possible, it was thought appropriate that incorruptible celestial bodies should govern the behavior of corruptible organic and inorganic bodies in the terrestrial region. This governance was accomplished by the continuous radiation of a variety of influences flowing unidirectionally from the heavens to the earth.

The skeletal frame of the world just described was frequently depicted in early printed editions of the fifteenth to seventeenth centuries. By the simplicity of its fundamental structure – represented by a series of nested concentric orbs embracing the terrestrial and celestial regions – this cosmic scheme satisfied European scholars psychologically and intellectually for some four hundred fifty years.

## The operational details

But if Western Europe was largely agreed on the macro structure of the world, it reached no consensus on the manner in which the great variety of specific cosmic activities occurred. Natural philosophers were in disagreement about many operational details, as is evident in their varied responses to the questions they regularly asked about the workings of the physical world. Medieval natural philosophy was constituted from the hundreds of questions that sought to cope with these operational details. But it was also something that came to transcend the sum total of the questions of which it was comprised. To characterize natural philosophy as it was understood and practiced in the Middle Ages, it is essential to explain its place in the knowledge schema of that period.

## WHAT IS NATURAL PHILOSOPHY?

Aristotle's division of the sciences provides the context for the place of natural philosophy within the framework of human knowledge. In his *Metaphysics*, Aristotle divides the sciences into *theoretical*, which is concerned with knowledge; *practical*, which treats of conduct; and *productive*, which deals with the making of useful objects. Aristotle further subdivides theoretical science into three parts: (1) theology, or metaphysics, as it was usually called, which considers things that exist separately from matter or body and are unchangeable – that is, spiritual substances and God; (2) mathematics, which also treats of things that are unchangeable, but only things that are abstracted from physical bodies and therefore have no separate existence, such as numbers and geometric figures; and (3) physics, which treats of things that not only have a separate existence but are changeable and possess an innate source of movement and rest. Physics was applicable to both animate and inanimate bodies. Aristotle's broadly conceived idea of physics is virtually equivalent to what is meant by natural philosophy, or natural science, as it was occasionally called. In the opening paragraph of his *Meteorology*, Aristotle explains what he understands by the study of nature, or natural philosophy.

Within this theoretical discipline he includes the study of the first causes of nature, change and motion in general, the motions of celestial bodies, the motions and transformations of the elements, generation and corruption, the phenomena in the upper region of the atmosphere right below the lunar sphere, and the study of animals and plants. In its broadest sense, then, natural philosophy in the Middle Ages was concerned with the study of bodies undergoing some kind of change. Or, as an anonymous fourteenth-century author put it approvingly, "the whole of movable being is the proper subject of natural philosophy."[3] The domain of natural philosophy thus was nothing less than the entire physical world, because motion and movable things occur everywhere in both the celestial and terrestrial regions, the two major parts into which Aristotle divided the cosmos.

For Aristotle, nature was comprised of bodies that are composed of matter, a passive principle, and form, an active principle. Natural philosophers studied the motions and changes of these bodies. As a union of matter and form, every body is affected by four basic causes: material, formal, efficient, and final. Acting simultaneously on all bodies, these four causes produce the never-ending sequence of cosmic effects. Mathematicians studied the same bodies as did natural philosophers, but from a radically different vantage point. They sought to abstract and study the geometric properties and features of material bodies and were therefore primarily concerned with the measurable and quantifiable aspects of bodies, not in the bodies themselves. Sciences that involved the application of mathematics to natural phenomena, such as optics, astronomy, and statics, were characterized as "middle sciences" (*scientiae mediae*) because they were assumed to lie between natural philosophy and pure mathematics.

During the Middle Ages, the relationship between natural philosophy and mathematics was occasionally, and perhaps even frequently, viewed quite differently. Around 1230, an anonymous author drew up a guide book for arts students at the University of Paris and divided natural philosophy into metaphysics, mathematics, and physics, thus assigning mathematics to natural philosophy. In this approach, natural philosophy has become equivalent to the whole of Aristotle's threefold division of theoretical knowledge. Nothing better illustrates the growing authority of natural philosophy. Some of those who did not accept this arrangement excluded the middle sciences from natural philosophy but continued to debate whether these sciences were closer to the latter or to mathematics, with partisans found on both sides. Where mathematics was applied to natural philosophy, the combination was thought of as independent of the middle sciences and was judged part of natural philosophy, as, for example, in the intension and remission of forms (see chapter 6).

In the form it took in the medieval universities, natural philosophy was a theoretical discipline studied by means of reason, analysis, and metaphysics. As it had been conceived by Aristotle and by those who wrote on the classification of the sciences, magic was excluded. Astrology played a role, but only because it had been an integral part of astronomy and also played a part in medical studies. If alchemy was considered, it was only because it was bound up with Aristotle's matter theory. Whatever their significance for the history of science, magic, astrology (especially as it pertained to human fate and fortune), alchemy, and other occult sciences were not officially taught in the natural philosophy curriculum of medieval universities, although this tells us little about the extent to which individual masters and students may have pursued these activities privately.

## THE QUESTIONS IN NATURAL PHILOSOPHY

A concrete sense of medieval natural philosophy requires knowledge of the kinds of questions that medieval natural philosophers asked about the world described by Aristotle and analyzed in his natural books. Earlier in this chapter, in tallying questions on five of Aristotle's natural books written by three different authors, we arrived at a total of 308 questions. If questions on Aristotle's *Metaphysics* and *Small Natural Books* are added to that total, the aggregate of questions might reach 600, or more. Whatever the approximate number, the totality of questions on these Aristotelian treatises constitutes the heart of medieval natural philosophy. It is essential, therefore, to convey what those questions were like. For convenience, I shall proceed in a cosmic order, citing first questions about the outermost parts of the celestial region, and what might lie beyond, and then proceeding downward to the lunar orb and then on to the upper terrestrial region and downward to the earth itself.[4]

The story really begins with questions about the status of the world as a created entity or as something that had no beginning and will have no end: or put in the form of a popular question, "Whether the universe could have existed from eternity." Numerous questions were but variations on this potentially hazardous theme. These included, for example, whether God could preserve the world through all eternity, and whether the world was something that was generable and corruptible, or was ungenerable and incorruptible. The possible existence of an eternal motion was yet another version of the basic question. By asking "whether the world was created," medieval natural philosophers were effectively inquiring about the eternity of the world, because a negative response would have implied a commitment to eternity. However, the question was usually posed from the standpoint of eternity rather than creation, because the questions were asserted within the context of one or another of Aristotle's natural books. By contrast, when theologians commented on Peter Lombard's *Sentences*, they regularly included questions about the creation, because the second book was devoted to it.

On the assumption that God had created the world, it became natural to inquire – especially after the Condemnation of 1277 – about what might lie beyond. In their consideration of extracosmic existence, natural philosophers wondered about the possible existence of other worlds, about the possible existence of an infinite void space spread out beyond our world, and about whether God is omnipresent in that infinite void space – indeed, whether God's omnipresence is co-extensive with that infinite void space. On the assumption that other worlds identical to ours did exist, natural philosophers asked whether the earth in one world would move naturally to the center of another world.

In accordance with their faith, however, scholastic authors devoted most of their analyses to a world that they believed was uniquely created. A major concern about that world was its state of perfection: did God make it perfect? What did it mean to characterize the world as "perfect"? And if it were perfect, could God have made it even more perfect? Although the actual size of our unique world did not materialize as a question, its finitude or infinitude attracted much attention. As if to test for the possibility of an infinite world, scholastic authors asked whether an infinite body could undergo circular or rectilinear motion. Because Aristotelian physics and cosmology would have been impossible if the world were physically infinite, negative replies to such questions led inevitably to the assumption of a finite world, which was in fact a cardinal principle of medieval natural philosophy.

The material composition of our finite world was a central theme. Did a fixed number of elements exist from which everything was made? Were there five elemental, or simple, bodies – the four traditional elements (earth, water, air, and fire) plus a fifth, an ether from which all celestial bodies were formed? Numerous questions were posed about alleged differences between celestial and terrestrial matter. Indeed, a perennial question asked whether the celestial, or supralunar, region, which was thought to be incorruptible, possessed matter in the same sense as did the terrestrial region, where change occurred incessantly. Another way of posing this question was to inquire whether celestial and terrestrial matter are of the same species, that is, whether they are essentially identical. Questions were also raised about the sameness or differences of the celestial orbs and planets. Natural philosophers frequently asked whether all celestial bodies – orbs, fixed stars, and planets – belonged to the same species, or whether each heavenly body constituted a unique species.

That the celestial region was filled with concentric spheres, or shells, each further subdivided into eccentric orbs, was axiomatic. The most popular cosmological question in the Middle Ages concerned the number of concentric spheres that were nested from the moon upward to the sphere of the fixed stars, and even beyond. The order that Ptolemy, the great Greek astronomer of the second century A.D., had given to the planets in his *Almagest* was usually the order accepted in the Middle Ages: Moon, Mercury, Venus, Sun, Mars, Jupiter, Saturn, and the sphere of the fixed stars. Disagreement persisted, however, in the arrangements of the movable orbs that were alleged to exist beyond the fixed stars, some of which had astronomical functions. Were all celestial orbs in motion, as Aristotle believed? Scholastics frequently asked whether an immobile sphere might lie beyond the mobile spheres, enclosing them as the container of the universe. Such an orb was usually assumed to exist and was known as the empyrean heaven, where the blessed and the elect of God live in eternal bliss, immersed in a dazzling light beside which our mundane light pales. Although the existence of the empyrean was assumed by almost all natural philosophers, questions were raised about it: is it a body? should it be called a heaven? how did it compare to other heavens?

By speaking of the waters above the firmament, the Bible provided the basis for two celestial orbs: the firmament and the crystalline heaven. Questions about those waters and the firmament on which they rested were usually posed by theologians in commentaries on the *Sentences* of Peter Lombard. Why are there waters above the firmament and what is their purpose? Do the waters have an orbicular shape? Does the crystalline heaven have the nature of water, or is it hard like ice? What is the firmament? Does it have a fiery nature? Many questions were raised about the properties associated with the celestial spheres and the planets. Were they incorruptible? Are planets and fixed stars spherical in shape? Can the heavens as a whole be characterized as heavy or light, or rare or dense? Indeed, if such properties existed in the heavens, did they differ from the heavy and light, and rare and dense, operating in the terrestrial region? That is, was the density of celestial bodies different from the density of terrestrial bodies? An important and difficult question asked whether orbs were distinct from each other and discontinuous, or whether the celestial region is one continuous body. A traditional problem, on which Aristotle gave no clear directions, concerned the animation of the heavens. Were they alive in some meaningful sense? Surprisingly, neither the Church fathers nor the Church adopted an official position on this important question, although the bishop of Paris regarded the attribution of life to celestial bodies as a dangerous idea and condemned it in 1277.

The celestial motions and their causes were a major preoccupation. Without exception, celestial motions were regarded as circular, uniform, and natural. Various arguments for these beliefs were given in questions that asked "whether circular motion is natural to the heavens" and "whether the sky is always moved regularly." Differences in speed among the orbs was also considered, as was the possibility that the same orb might move with several simultaneous motions.

As for the causes of celestial motions, natural philosophers suggested external and internal possibilities when they asked "whether the heavens [that is, the orbs of the heavens] are moved by intelligences, or intrinsically by a proper form or nature?" Whatever the cause, those same scholars also inquired whether the motive causes might become fatigued and cease producing uniform motions.

Because of its role in the creation account and its dramatic manifestation in the heavens, celestial light was a fundamental theme. Theologians were much interested in whether God had created light on the first day and, if so, what its nature might be. Of major interest to all natural philosophers, however, was the source of planetary and stellar light. Was each celestial body the source of its own light; or, were the planets essentially devoid of light and dependent on the sun for all the light they received?

On the assumption that "the more powerful and superior should influence the less powerful and inferior,"[5] as Saint Bonaventure expressed it, scholars frequently asked whether the celestial region influenced things in the inferior world, below the sphere of the moon. They inquired about the influence of the various components of the celestial region – that is, what was the influence of the planets? did each planet exert a different influence? what was the influence of light? of motion? Indeed, on the assumption that celestial motions exercised a pervasive influence on inferior bodies, scholastics sometimes asked "whether all motions and actions of inferior bodies would cease if the celestial motions ceased."

Questions about activities in the terrestrial region and about the four elements appear primarily in questions treatises on Aristotle's *On Generation and Corruption* and, to a lesser extent, in those on his *Meteorology* and *On the Heavens*. Natural philosophers asked about the location of the elements, as well as about their magnitudes and shapes. They were also concerned about whether elements retained their identity in a compound; whether one element could be generated directly from another; and whether an element could exist in nature in a pure state. Questions were also raised about individual elements. Is fire hot and dry? Does it move with circular motion in the region just below the moon? Is light the form of fire? They inquired whether air is naturally hot and humid and then, surprisingly, asked whether the middle region of air is always cold. Because comets, meteors, and the Milky Way were assumed to be sublunar phenomena, questions were regularly asked about them. For example, do comets have a celestial nature, and do they foretell wars, plagues, and the death of rulers?

The motions of elements and compounds prompted numerous questions. Natural philosophers determined that there was a predominant element in a compound body and then asked whether that predominant element determined the direction of the body's motion. The ways in which heavy bodies move with natural or violent motions under a variety of conditions were usually discussed in questions on the *Physics* and, to a lesser extent, on *On the Heavens*. For example, the natural philosophers asked whether bodies that move upward or downward have internal resistances; whether a resistant medium, such as air or water, is essential for motion to occur. It then seemed natural to inquire whether a separate, extended vacuum might exist, and, if so, could bodies move through it with finite speeds? Although Galileo, Descartes, and Newton made some of their greatest contributions to science in the kinematics and dynamics of motion, many of the questions proposed in the Middle Ages form part of the history of those solutions. For example, it was common to inquire whether the mover and the thing moved are conjoined; whether a moment of rest intervenes between the upward and downward trajectories of a violent motion; and, one of the most important questions, whether, after leaving the hand of the projector, a projectile is moved by the ambient air or by an impressed force, or impetus. Impetus was invoked to explain the natural acceleration of falling bodies in questions that asked "whether natural motion is swifter in the end than at the beginning" or "whether after departure from a projector, a stone that is projected, or an arrow that is shot from a bow, and so on in similar cases, are moved by an internal principle or an external principle."

Medieval natural philosophers customarily considered a variety of questions about the earth as a whole. One of these concerned its sphericity, especially the manner in which the earth's mountains and irregularities could be reconciled with its sphericity. There were also questions about the earth's relative size. For example, is the earth like a point in comparison to the heavens; and is the magnitude, or size, of the earth smaller than that of certain planets? Other noteworthy questions concerned the earth's location (was it fixed in the center of the world?); the distribution of its matter (did the earth have the same center of gravity and magnitude?); and its status at the center of the universe (was the whole earth at rest in the center of the universe, or did it rotate on its axis?). Natural philosophers also regularly inquired whether the whole earth was habitable.

## THE TECHNIQUES AND METHODOLOGIES OF
## NATURAL PHILOSOPHY

Medieval natural philosophy consisted of the kinds of questions and topics just described. It was concerned with hundreds of questions that ranged over the world from the outermost celestial orb to the bowels of the earth. The methodology for treating questions in natural philosophy was at least twofold. One kind concerned the abstract analysis of science and sought to determine what a demonstration is in science and the nature of causal relations. The other involved techniques that were used to support or reinforce arguments.

## Abstract methodology

During the Middle Ages, the ideal of science was demonstration by means of a syllogism. Although it was based on Aristotle's *Posterior Analytics*, notions of what it meant varied considerably. Whatever was thought about the means of achieving demonstrative knowledge, most scholastics agreed that, insofar as it was possible, the goal of science and natural philosophy was to demonstrate truths about the world. Some theologian-natural philosophers were disturbed by the thought that the certainty attainable by Aristotelian demonstrative science would rival that of faith, and perhaps subvert it. To counter this unpleasant possibility, some theologian-natural philosophers created doubts about the certainty of Aristotelian demonstrative science by invoking the powerful doctrine of God's absolute power, which, as we saw, was an important factor in the Condemnation of 1277.

The most significant player in this drama was William of Ockham (ca. 1285–1349). A gifted logician and philosopher, Ockham was also an outstanding theologian. In his view, the world was utterly dependent on the unfathomable will of God, who, by his absolute power could have made things other than they are. From this it followed that all things in existence are contingent – that is, they could have been made otherwise, or not at all. As a completely free agent, God can do anything that does not involve a logical contradiction. Whatever he could create through secondary or natural causes, he could also produce and conserve directly, or concomitantly with secondary or natural causes. So great is God's power that he could, if he wished, create an accident without its substance, or a substance without its accidents; or produce matter without form, or form without matter. From these strictly theological considerations, which reflect the theological spirit that produced the Condemnation of 1277, Ockham derived an epistemology that has been characterized as radical empiricism.

The major feature of Ockham's empiricism is the conviction that all knowledge is gained by experience through "intuitive cognition," an expression Ockham adopted from Duns Scotus. By this Ockham meant that objects external to the mind, as well as personal mental states, are grasped directly and immediately. Direct perceptions of this kind permit one to know whether or not something exists. No demonstration is required, and none can be produced, to show the existence of anything apprehended in this manner. Indeed, even an absent or inaccessible object might produce an intuitive cognition, because God himself might choose to supply the cause of the cognition directly, rather than operate in the customary manner through a secondary cause. In either event, our experience of that object would be the same. Indeed, God could also cause us *to believe* in the existence of an object that does not actually exist, but he cannot make us have evident knowledge that it exists. That is, God can cause us to have a belief that an object exists, but he cannot make us know that it really exists. That would be a contradiction, because the object has been assumed not to exist. For Ockham, then, psychological certitude was indistinguishable from certitude based on "objective" evidence acquired through the senses.

By denying necessary connections between contingent things, Ockham was led to an examination of causal relations. In his *Commentary on the Sentences*, he argued that something could be identified as an immediate cause when the effect it produces occurs in its presence and – all other things being equal – fails to occur in its absence. Only by experience, however, not a priori reasoning, could sequences of events meeting the conditions described here be justifiably characterized as causally related, as, for example, when we detect that fire is the cause of combustion in cloth. Because Ockham had shown that the existence of one thing did not necessarily imply the existence of another thing, a priori reasoning plays no role, as it did in earlier discussions of causality. Even experience did not guarantee certainty in determining causal relations – God might have dispensed with the secondary cause and set fire to the cloth directly. Even under ideal conditions for observing repeated sequences of events, it would be impossible to identify the particular causal agent with certainty. Thus did Ockham appear to undermine the Aristotelian sense of definitely knowable and necessary cause–effect relationships that had been so widely accepted in the thirteenth century.

Although he had his greatest impact in the fourteenth century, Ockham exerted a pervasive influence well beyond it. Some who seem to have been influenced by him sought to abandon demonstrative science and rely on probable arguments. The theologians Nicholas of Autrecourt (b. ca. 1300–d. after 1350) and Pierre d'Ailly (1350–1420) and others in the fourteenth century sought to construct alternatives to Aristotle by arguing that probable solutions as good as Aristotle's were possible on most problems and should be used. They stressed that many arguments were not demonstrative and therefore could only be probable. Blasius of Parma (ca. 1345–1416), a teacher of mathematics and natural philosophy at various Italian universities, contrasted mathematics, which is a demonstrative science, with natural philosophy, a discipline that he believed was concerned with things not subject to demonstration. But even in the thirteenth century, Robert Grosseteste had argued that demonstrations in physics, or natural philosophy, were only probable, as contrasted with mathematical demonstrations, which were certain. Roger Bacon insisted that in natural philosophy experience had to confirm demonstration. "Therefore, reasoning does not suffice," argued Bacon, "but experience does." He concluded that "What Aristotle says therefore to the effect that the demonstration is a syllogism that makes us know, is to be understood if the experience of it accompanies the demonstration, and is not to be understood of the bare demonstration."[6] With regard to level of certitude, it was generally agreed that natural philosophy delivered less of it than mathematics, which provided the paradigm for certain demonstration.

Although demonstrative science was discussed and considered important, it played only a small role in the usual problems that were treated in regular questions treatises. Probable solutions were also rare. In fact, abstract methodology of the kind found in Aristotle's *Posterior Analytics* was uncommon in the treatment of real problems. In this, medieval natural philosophers differed little from Aristotle, who also had found few occasions to apply his scientific methodology to real physical problems.

### Methodologies that were actually used

When discussing specific questions about the natural world, scholastic natural philosophers were likely to follow the more practical methodology devised by John Buridan than that of Ockham and his followers. Because natural philosophers were fully aware that they lived in a time when reasoned arguments about the physical world had to be compatible with theological and doctrinal concepts that were rooted in over a thousand years of Christian history, they developed certain attitudes about their conclusions and interpretations. To allow for the possibility of divine intervention and to cope with nature's sometimes unpredictable actions, medieval natural philosophers developed some basic strategies and approaches that were sometimes made explicit but were often implied. Few made use of these better than John Buridan, perhaps the most brilliant arts master of the Middle Ages.

A confirmed empiricist, Buridan believed that humans could understand nature's operations in terms of cause and effect. He also accepted the truths of revelation as absolute, and he had no overt problems with his faith. Although he acknowledged that God could do anything that was considered impossible by natural means, Buridan was not attracted to the physics and cosmology of "what God might have done." God's uncontested power to do anything whatever short of a logical contradiction was not to be construed as implying that he had actually done so. Nevertheless, it was intellectually unsettling to acknowledge that cause-and-effect relationships might not always operate because God could choose to intervene miraculously and drastically alter an otherwise unalterable cause-effect relationship, for example, by making fire cold; or, that what appears to be an instance of a fire burning a log naturally, might actually be an instance in which God directly causes the log to burn, with the fire playing no role. Even without divine intervention, effects in nature sometimes failed to occur or were grossly distorted.

Under these uncertain circumstances, was natural truth attainable? In a question that inquired "whether the grasp of truth is possible for us"[7] Buridan followed a tradition that had already been underway. He argued that with regard to natural science, truth is attainable, provided that "a common course of nature (*communis cursus nature*) obtains in natural things, and in this way it is evident to us that fire is warm and that the heaven moves, although the contrary is possible by God's power."[8] As Bert Hansen has aptly explained: "Phenomena regarded as natural in the Middle Ages were only those which occur most of the time, in nature's 'habit' or usual course. The law of nature within the Aristotelian conceptual framework was not one of rigid necessity, but simply that of the usual or ordinary occurrence."[9] Defects occur in human births, but are regarded as anomalous. Natural causes usually produce their intended effects.

Buridan was convinced that "for us the comprehension of truth with certitude is possible."[10] The "certitude" Buridan had in mind consisted of indemonstrable principles that he regarded as the basis of natural science. Our belief that all fire is warm and that the heaven moves is based on inductive generalizations – or, as Buridan expressed it, "they are accepted because they have been observed to be true in many instances, and to be false in none."[11] Such truths are necessarily conditional because they are predicated on the assumption of a "common course of nature." They are also indemonstrable principles. God's possible intervention in the causal order is thus rendered irrelevant. For although God could alter the course of natural events, Buridan insists that "in natural philosophy, we ought to accept actions and dependencies as if they always proceed in a natural way."[12] Neither the occurrence of miracles nor anomalous chance events affect the validity of natural science.

Inductive generalization was a powerful tool in medieval natural philosophy. An indemonstrable principle could be constructed on the basis of one or two positive examples and no counterinstances. Additional instances would only deepen confidence in the principle but did not increase its validity.

Scholastic natural philosophers employed a number of other methodological techniques to strengthen or justify their arguments, or to increase the cumulative impact of their conclusions. Perhaps the most widely used was the principle of simplicity. In its most basic form, it derives from Aristotle, who declared, at least four times in his biological works, that nature does nothing superfluous or in vain.[13] On a number of occasions, Aristotle applied the principle, as when he declared that it was unnecessary to "assert an infinity of elements, since the hypothesis of a *finite* number will give identical results."[14]

During the Middle Ages, the principle of simplicity was frequently employed, with various twists. In its broadest sense, we know the principle as "Ockham's razor," named after William of Ockham. Although, as we shall see, Ockham intended his principle of simplicity, or parsimony, as it has been frequently called, to apply only to thoughts and not to things, he gave various versions of it that were equally applicable to either. Thus Ockham declared that "what can be done with fewer [means] is done in vain with many," and "plurality is not to be assumed without necessity."[15] Another famous variant, falsely attributed to Ockham, declared that "entities should not be multiplied beyond necessity" ("Entia non sunt multiplicanda praeter [or *sine*] necessitate").[16]

John Buridan applied yet another version of the simplicity principle to the earth's possible axial rotation. After first proclaiming an easily recognizable version of the principle – "it is better to save the appearances by fewer than by more [assumptions]" – Buridan added that it is also "better to save the appearances by an easier path than by a more difficult path."[17] To illustrate the claim, Buridan argued that it would be better – that is, simpler – to assume that the relatively small earth rotates daily on its axis than to assume that the vastly larger celestial orbs move with an enormously greater speed to produce a daily rotation of the heavens. Although Buridan did not follow his own advice – indeed, the principle of simplicity was often ignored when it was found inconvenient or when other seemingly good arguments vitiated it – the argument was one that Copernicus and Galileo found attractive.

In a world created by a God who had, by common agreement, the absolute power to do as he pleased, the principle of simplicity had its limitations, as William of Ockham observed. If nature always acted in the simplest and easiest way, this would imply that God had created a world in which all actions were accomplished in the simplest manner. But we cannot know whether God chose such a path. By virtue of his absolute power, God could have created a world in which things are complex and difficult rather than simple and easy or a world in which some things are complex and some simple. Ockham was apparently convinced that "God does many things by means of more which He could have done by means of fewer simply because He wishes it. No other cause [sc. of His action] must be sought for and from the very fact that God wishes, He wishes in a suitable way, and not vainly."[18] Although he thought that the principle of simplicity could not be applied with any degree of confidence to things in nature, Ockham was convinced that it could be applied to philosophical thought. We ought to explain our thoughts about things in the simplest possible manner and avoid the needless multiplication of explanatory entities.

In addition to simplicity, scholastic natural philosophers also used arguments from nobility and hierarchy. The idea that some things are better than others and that generally there is a gradation of goodness and virtue in nature was universally held. Aristotle incorporated this belief into his natural philosophy in the form of a ladder of nature. In the terrestrial region, inanimate objects were located at the lowest rung of the ladder, followed by plants, animals (including those thought to be spontaneously generated from mud and slime), and humans. Because the terrestrial region was a realm of incessant change, the celestial region beyond, where changes of substance, quantity, and quality were nonexistent, was assumed to be incomparably better and nobler than all things except human life and the immortal souls associated with it. Generally, things were considered nobler, and therefore "better," the greater their distance from the earth. Mars was nobler than the Sun because it was more distant from the earth; and Jupiter was nobler than Mars for the same reason, and so on. Like many similar medieval principles, it was violated as occasion demanded. Nicole Oresme, however, rejected the principle itself, arguing that "the perfection of the heavenly spheres does not depend upon the order of their relative position as to whether one is higher than another." For Oresme, the Sun, which occupies the middle position among celestial bodies, "is the most noble body in the heavens and is more perfect than Saturn or Jupiter or Mars, which are all higher up than the sun."[19] As a scholar who denied the principle of ascending nobility, Oresme was a rare exception.

The concept of nobility was even applied to rest and motion. Here there was no unanimity of opinion. Indeed, John Buridan regarded the rest acquired after a body reached its natural place as superior to the motion that brought it to its natural place. In the celestial region, however, he regarded motion as nobler than rest because celestial bodies are always in their natural places and have no other goal than to move in those places with their natural, circular motions. Scholastic authors seem to have assigned greater nobility to rest or motion largely on the basis of the needs of the particular argument in which they were engaged.

In chapter 5, we described a powerful analytic tool that involved the use of the imagination, where conditions were imagined that were impossible in Aristotle's natural philosophy and from which consequences were derived. Natural philosophers actually devised an expression to epitomize this approach when they referred to such counterfactuals as *secundum imaginationem*, that is, "according to the imagination." The Condemnation of 1277 played a significant role in generating counterfactuals. Many of the condemned articles compelled the acknowledgment of God's absolute power to do whatever he pleased short of a logical contradiction. Examples of counterfactuals that were derived from the Condemnation of 1277 include the possibility of other worlds, vacua within and beyond the world, and the possibility that God might move our world with a rectilinear motion. In each of these examples, medieval natural philosophers sought to derive consequences within an assumed framework of Aristotelian physics, even though what they initially assumed was impossible in Aristotle's system. What emerged was a series of interesting speculations, or, as we might say today, thought experiments, in which certain Aristotelian principles were challenged and, to some extent, subverted. For example, the idea of the existence of other worlds was shown to be an intelligible concept, even though Aristotle had argued that other worlds were impossible.

The methodologies I have described here – the common course of nature, inductive generalization, simplicity, nobility, and hierarchy – were elaborations on the Greco-Arabic inheritance. Some of them – especially simplicity, nobility, and hierarchy – were based on a priori, metaphysical assumptions. All, however, were tools meant to enhance and reinforce arguments. Other methodological instrumentalities might be mentioned, but those included here were among the most popular and useful.

## THE ROLE OF MATHEMATICS IN NATURAL PHILOSOPHY

From the preceding discussion about methodology in natural philosophy, we may plausibly infer that natural philosophers did not regard natural philosophy as generically different from the exact sciences. Indeed for many natural philosophers, the exact sciences, or middle sciences, such as astronomy and optics, were but the more mathematical aspects of natural philosophy. Mathematics and natural philosophy, however, had an even closer relationship than their formal connections through the middle sciences would indicate. The application of mathematics to problems of natural philosophy was fairly common in the Middle Ages. As we saw in chapter 6, Aristotle's widely used description of motion was transformed in the fourteenth century by Thomas Bradwardine, who abandoned Aristotle's version, which was expressed in terms of arithmetic proportionality (in the form $V \propto F/R$) and replaced it with a function expressed in terms of geometric proportionality. Bradwardine's *Treatise on Proportions*, where he made this significant transformation, is rightly regarded as the medieval equivalent of a treatise in mathematical physics.

In other areas of medieval thought, mathematics was a recognized tool of analysis, as in the doctrine of the intension and remission of forms, and in a wide variety of problems that involved infinite processes. In chapter 6, we saw that the intension and remission of forms or qualities developed from two initially distinct problems, one in natural philosophy concerned with how qualities vary (based on chapter 8 of Aristotle's *Categories*) and the other in theology involving the possible variation of grace and charity in humans (Peter Lombard's *Sentences*, book 1, distinction 17). From these beginnings, a rationale was eventually provided for the quantitative treatment of a whole spectrum of qualities. A quality – say, redness or hotness – was believed augmentable and diminishable in the same manner as, for example, weights, or extended magnitudes. Medieval natural philosophers believed that three degrees of redness could be added to two degrees of redness to produce five degrees of redness. In the last three-quarters of the fourteenth century, it was assumed that identical qualitative parts could be added and subtracted. Thus one could treat qualitative changes mathematically. Attention was focused on how best to represent the different possible modes in which qualities were thought to add or lose parts. Some – for example, those at Merton College, Oxford – chose to represent these qualitative alterations arithmetically, whereas others, especially Nicole Oresme, somewhat later on, used geometrical figures to achieve similar results. By analogy with the alteration of qualities, scholastic natural philosophers came to treat motion as if it were just another quality. Because of this association of motion with qualities, significant theorems about motion were derived that forever link the medieval doctrine of intension and remission of forms with Galileo's contributions in the seventeenth century. As interest in the mathematical treatment of qualities increased, there was a corresponding loss of interest in the theological and metaphysical aspects of qualitative change that had been paramount in the early stages of development. (For more on intension and remission of forms, see chapter 6.)

Those who wrote on this subject between the late fourteenth and sixteenth centuries came to be called collectively "calculators" (*calculatores*), an apt term that signified their effort to measure (by mathematical techniques) the increase and decrease in intensity of qualities, as if the latter were extensive magnitudes. The original calculators were fellows of Merton College during the 1330s and 1340s. The group included Walter Burley, John Dumbleton, William Heytesbury, Roger Swineshead, Richard Swineshead, and Thomas Bradwardine. Some of them – Burley, Heytesbury, and Bradwardine – eventually became theologians, although they did much, if not most, of their work in natural philosophy while they were still arts masters. The ideas of the Oxford calculators were disseminated to Italy in the fifteenth and early sixteenth centuries. These ideas reached Paris during the second half of the fourteenth century and were revitalized there during the first half of the sixteenth century. Gottfried Leibniz paid high tribute to Richard Swineshead by having the 1520 Venice edition of the latter's *Liber calculationum* transcribed. It seems that

Leibniz regarded Swineshead's application of mathematics to natural philosophy as a major achievement.

Calculators were also quite interested in what happens in the first and last instants of potentially infinite processes. Here was yet another significant attempt to treat problems in natural philosophy mathematically. A problem that lent itself to such analysis involved the separation of two plane surfaces (or plates) or the approach of two such surfaces before contact. Would the moment of separation produce a vacuum? At first glance, it appears that if two plane surfaces, initially in uniform, mutual contact and with no material medium intervening, were separated in such a manner that they remained parallel after separation, a momentary vacuum would necessarily form. Why? Because immediately after separation, air would rush in to fill the intervening space, but before it could travel successively from the outer perimeters of the plane surfaces to the innermost parts, a brief time interval must elapse during which a momentary vacuum would exist around the center. Because medieval natural philosophers unanimously accepted the dictum that "Nature abhors a vacuum," they could not believe that nature would allow even a momentary, small vacuum. Thus they proposed various solutions to demonstrate that a vacuum would not form.

Blasius of Parma agreed with his colleagues. In a work titled *Question on the Contact of Hard Bodies* (*Questio de tactu corporum durorum*), he insisted that no vacuum can form between the two plane surfaces. Blasius described a situation in which two perfectly circular plates first come into contact without forming a vacuum and then explained why a vacuum will not form when the plates are separated after being in direct and perfect contact.[20] In the first case, when two separate plates come into direct contact, Blasius assumed that as the surfaces approach but do not yet touch, the air between the plates becomes rarer and rarer as more and more of it escapes from between the plates. Although it becomes rarer and rarer, the intervening air does not part to allow formation of a vacuum. Blasius concluded that there is no last assignable instant in which rarefaction ceases prior to contact of the surfaces.

Actual contact, however, must be interpreted as the last move in the completion of the process of motion for both plates, but as lying outside that process. Thus actual contact is construed as lying outside the process of motion toward contact. In effect, there is no last instant in which the surfaces are separated, although there is a first instant in which they are in contact. But if there is no last instant in which the surfaces are separated, there can be no last instant in which the rarefied air departs from the intervening space. Consequently, no vacuum can occur just before contact.

With the two circular surfaces in direct contact, Blasius explained next how their subsequent separation will also fail to produce a vacuum. To demonstrate this, he has to show that immediately after separation air completely fills the space between the surfaces. He does this by assuming that a first moment of separation cannot be determined. Thus there is a last moment, or instant, of contact, but no first instant of separation. For if a first instant of separation exists, there must also be a minimum distance of separation. However, given any initial distance of separation, one can always argue that at some previous time the surfaces must have been separated by half that distance, and so on. Hence there can be no initial distance of separation and consequently no first instant of separation. But if there can be no initial distance of separation, and therefore no first instant of separation, it follows that for any moment chosen after the separation of the circular plates, air will fully occupy the intervening space associated with that particular moment. Therefore no vacuum can occur. Thus did Blasius of Parma make rather ingenious use of the doctrine of first and last instants.

In the late Middle Ages, mathematics was recognized as important for natural philosophy. In his *Treatise on the Continuum* (*Tractatus de continuo*), Thomas Bradwardine emphasized the importance of applying mathematics to natural philosophy, especially to problems involving the nature of continua in objects and things, where it was necessary to determine whether continua were infinitely divisible or were composed of indivisible units or atoms. Nicole Oresme, Blasius of Parma, the calculators, and numerous others used mathematical concepts to elucidate natural philosophy. Should we, therefore, view the medieval application of mathematics to natural philosophy as the real beginning of mathematical physics? Was the application of mathematics to physics in the seventeenth century just a continuation and extension of medieval practice? Although there may be some connections, the two approaches toward mathematical physics differed considerably, if not radically. Those who were instrumental in producing the Scientific Revolution, such as Galileo, Descartes, Kepler, and Newton, sought to apply mathematics to real problems in the physical world. By contrast, medieval applications of mathematics to natural philosophy were usually of a hypothetical character divorced from empirical investigation. They were all too often purely formal exercises based on arbitrary assumptions and dependent on logical arguments. Rarely did medieval natural philosophers claim correspondence between their conclusions and the "real" world. Indeed, they had virtually no interest in testing their hypothetical conclusions against that world. It was "natural philosophy without nature," as John Murdoch has so aptly characterized it.[21] However, except for counterfactuals that considered "natural impossibilities," or speculations about hypothetical possibilities that may or may not have involved the application of mathematics, most of the non-mathematical questions posed by medieval natural philosophers on the natural books of Aristotle were about the real world. The replies were also thought to be real solutions about that world.

## THE USE OF NATURAL PHILOSOPHY IN OTHER DISCIPLINES

Natural philosophy was so broad that it was inevitable that it would affect other disciplines, and even pervade them. Strong connections are apparent from the very fact that knowledge of natural philosophy was regarded as a prerequisite for entry into other, "higher" disciplines.

### Theology

During the late Middle Ages, natural philosophy, along with the mathematical concepts that were an integral part of it, were frequently, if not regularly, applied to theological problems in the *Sentences* (*Sententiae*, or opinions) of Peter Lombard (d. ca. 1160). Divided into four books devoted to, respectively, God, the creation, the incarnation, and the sacraments, the *Sentences* was for more than four centuries the standard text on which all theological students were required to lecture and comment. The second book, devoted to the creation, afforded ample opportunity to apply natural philosophy to a variety of themes on the six days of creation. Natural philosophy had an equally significant impact on theology in problems that involved God's relationship to the world and its creatures. The injection of natural philosophy, mathematics, and logic into commentaries on Peter Lombard's *Sentences* grew to such proportions that in 1366 the University of Paris decreed that, except where necessary, those commenting upon that work should avoid the introduction of logical or philosophical material into the treatment of the questions (such restrictions were never applied at Oxford University). Long before that, Pope John XXII (1316–1334) had reproached the theological masters at Paris for treating philosophical questions and subtleties. In 1346, Pope Clement VI (1342–1352) castigated the theologians at the University of Paris for ignoring study of the Bible in favor of philosophical problems and disputes. Despite such appeals, scholastic commentators apparently found it "necessary" to introduce such matters frequently and extensively. In a telling remark, the theologian-natural philosopher John Major (1469–1550) declared in his introduction to the second book of his own commentary on the *Sentences*, that "for some two centuries now, theologians have not feared to work into their writings questions which are purely physical, metaphysical, and sometimes purely mathematical."[22]

Theologian-natural philosophers were not only concerned with the creation. They frequently applied themes, techniques, and ideas from natural philosophy to problems involving God's omnipresence, omnipotence, and infinity, as well as to his relations to the beings of his own creation and to comparisons between created species. Theologians even found occasions to use concepts of natural philosophy to explain various aspects of the mysteries of the mass, or Eucharist. The application of these ideas from natural philosophy, most of which conflicted with theological dogmas, to the Eucharist has been aptly called the "Physics of the Eucharist." In 1215, the Lateran Council, convoked by Pope Innocent III, established the doctrine of the Eucharist. In the mass, it was assumed that the bread and wine were miraculously transformed into the body and blood of Christ. After the transformation, the body, or substance, of Christ replaced the substance of the bread. However, the visible properties, or accidents, of the bread remained. But in what do those accidents inhere? They are not in Christ, because the body of Christ cannot be identified or associated with the bread. They can no longer be in the bread, because the bread has been transformed into the body of Christ. Where are they? It was concluded that the remaining visible accidents of the bread do not inhere in any substance. Although this was a miraculous occurrence, it clashed directly with Aristotle's natural philosophy, which assumed that every accident or property inhered in a substance. Indeed, it was naturally impossible for an accident to exist independent of a substance. The doctrine of transubstantiation, or the doctrine of the mass or Eucharist, posed numerous problems for Aristotelian natural philosophy and for the theologians who used it. Ironically, the Lateran Council took place in the early thirteenth century, just when Aristotle's natural books were emerging as the dominant intellectual force in university life.

The Church and its theologians understood that Christ was present substantially and accidentally in the Eucharist. Indeed, Christ was assumed to be wholly in every part of the host, an assumption deemed necessary in order to avoid the unacceptable consequence that, when the host was split apart upon being eaten, Christ would also be broken into parts. But if the whole of Christ is present in every part of the host, however small, his body cannot be an extended quantity within the host. One problem was to determine how Christ's unextended presence could be reconciled with Aristotle's doctrine of quantity, which was always associated with extension. The Eucharist raised other serious dilemmas for natural philosophy, especially concerning Aristotle's doctrines of change and of place. For example, Christ was assumed to be in place in the host differently than a physical body would occupy a place. The latter is co-extensive with its place in length, depth, and width. But, as a spiritual entity, Christ was in the host "definitively," that is, he was not necessarily coterminous with its boundaries, but was located somewhere within. In coping with such problems, some scholastic theologians

– Thomas Aquinas is a good example – sublimated natural philosophy to the needs of theology, stretching it where necessary. Others, like William of Ockham, used natural philosophy in theology where they thought it could be legitimately applied, without distorting it for the requirements of theology. Where natural philosophy seemed inapplicable to theology, they resorted to revelation, dogma, and God's absolute power.

Theologians were obviously not content simply to assert the doctrine of the Eucharist dogmatically. They sought rather to explain its many problems by utilizing natural philosophy. The "Physics of the Eucharist" was not confined to the Middle Ages. It was still an issue in the seventeenth and eighteenth centuries, when John Locke (1632–1704) and Gottfried Leibniz (1646–1716) discussed it, with Leibniz even involving Newton's theory of gravitation.

Theologians frequently used mathematical concepts developed within natural philosophy. These included applications of proportionality theory to physical problems, the nature of the mathematical continuum, convergent and divergent infinite series, the infinitely large and small, potential and actual infinites, and the determination of limits involving first and last instants of infinite processes. The determination of limits, for example, was found useful in problems concerning free will, merit, and sin. In the fourteenth century, the English theologian Robert Holkot conceived the following dilemma: either we place limits on free will, or we concede that God might not always be able to reward the meritorious and punish the sinful. He imagined a situation in which a man is alternately meritorious and sinful during the final hour of his life. Thus he is meritorious during the first proportional part of his final hour and sinful in the second proportional part; he is again meritorious in the third proportional part, and again sinful in the fourth proportional part, and so on through the infinite series of decreasing proportional parts up to the last instant, when death occurs. Because the instant of death cannot form part of the infinite series of decreasing proportional parts of the man's final hour, it follows that there is no last instant of his life and, therefore, no last instant in which he could be either meritorious or sinful. Under these circumstances, God cannot know whether to reward or punish this man in the afterlife, an obviously unacceptable consequence of the doctrine of free will. One could only conclude that free will cannot be extended to every imaginable sequence and pattern of choices, a point that Holkot buttressed with eight additional continuum arguments.

Dilemmas associated with the actual infinite posed perplexing theological problems. At the end of the sixteenth century, Jesuits who taught at the University of Coimbra in Portugal, and who are usually referred to as the "Coimbra Jesuits," inquired (in their *Commentary on Aristotle's Physics*) "whether by his divine power God could produce an actual infinite." In responding to this frequently posed medieval question, they explained that "in this weighty and difficult controversy, the negative part is to be preferred [namely that God cannot create an actual infinite], both because it advances the better arguments and because the better-known philosophers have embraced it." A few lines later, however, they added an important qualification by explaining that "since the affirmative part [that God can create an actual infinite] also has its probability and not unworthy supporters, and especially because by no [single] argument can it be plainly refuted, we explain the arguments of each side, so that anyone can support any side he wishes."[23]

The language of mathematics and measurement was embedded into natural philosophy during the fourteenth century. Theologians eagerly appropriated this new and exciting language. Not only did they use it in the difficult domain of free will and sin, but they applied it to a variety of other problems. They employed it to describe the manner in which spiritual entities could vary in intensity, using for this purpose the peculiarly medieval doctrine of "intension and remission of forms or qualities" (occasionally known as "the configuration of qualities"; see chapter 6). These mathematical concepts were also useful in problems concerned with infinites, as, for example, speculations about God's infinite attributes (namely, his power, presence, and essence); the kinds of infinites God might possibly create; the infinite distances that separate him from his creatures, a problem relevant to the widely discussed concept of the perfection of species; the possible eternity of the world; whether God could improve upon something he had already made, especially whether he could make better and better successive worlds without end; and whether he could create an ultimate, best possible world.

The behavior of angels also proved a fertile area for the application of various aspects of natural philosophy. Theologians were fascinated with the modes of angelic existence. For example, they asked if an angel occupies a place. If so, they might then inquire whether an angel could be in two places simultaneously; whether two angels could occupy the same place simultaneously; and whether an angel moves between two separate places with a finite or an instantaneous speed. Responses to all these questions were given in terms of concepts that had been developed in natural philosophy in discussions about the motion of material bodies. Angelic motion became one of the most popular contexts for the intense medieval debate about the nature of the continuum: whether it consisted of parts that are infinitely divisible, or was composed of indivisible, mathematical atoms that could be either finite or infinite in number.

The importation into theology of concepts and techniques from natural philosophy, and especially from mathematically oriented topics in natural philosophy, led theologians to express their problems in a logico-mathematical format that was essentially hypothetical and speculative, or "according to the imagination" (*secundum imaginationem*) as they would have said in the Middle Ages. Why this occurred is not obvious. Perhaps it derived from the widespread conviction among theologians that God's nature and the motives for his actions were not directly knowable by human reason and experience, thus making it convenient to express theological problems in hypothetical form. The common educational background of theological students and masters may also have played a role. They had all been heavily exposed to natural philosophy and logic and, to a lesser extent, to geometry as well. With this kind of background, many theologians were attracted to the techniques of the calculators and sought to couch hypothetical theological problems in the logico-mathematical format that had been incorporated into natural philosophy during the first thirty or forty years of the fourteenth century.

Whatever the reasons for the hypothetical and quantitative style, it resulted in a major change in the techniques of theology. Solutions to many theological problems were sought by various quasi-logico-mathematical measuring techniques drawn from natural philosophy. Traditional theological questions were often recast in a quantitative mold that allowed the easy application of logical and mathematical analysis. And yet this massive influx of quantitative apparatus appears to have had little if any impact on the content of theology, even as it transformed its methodology. As a result, the fourteenth century, and the late Middle Ages, was an extraordinary period in the history of the relations between science and theology in the Western world.

### Medicine

Almost from its beginnings in ancient Greece, with the works of Hippocrates, medicine has been intertwined with natural philosophy. So extensive was this interrelationship that the Hippocratic author of *On Ancient Medicine* sought to counter its influence by arguing that philosophy is useless for the study of medicine. Indeed, he argued that only through a study of medicine can we know nature. But his was not the dominant voice. Large doses of natural philosophy were injected into Hippocratic medical treatises. Aristotle, himself the son of a physician, thought that the study of medicine was essential for the study of nature. In his *On the Senses* (*De sensu*), he urged students of nature "to inquire into the first principles of health and disease,"[24] and in *On Youth and Old Age* (*De juventute et senectute*), he declared that "those physicians who are cultivated and learned make some mention of natural science, and claim to derive their principles from it, while the most accomplished investigators into nature generally push their studies so far as to conclude with an account of medical principles."[25] For Aristotle, natural philosophy and medicine were intimately related. The greatest of all

Greek physicians, Galen, was also a philosopher and found numerous occasions to intrude natural philosophy into his writings. His most significant Islamic commentator, Avicenna, who also commented on Aristotle's natural philosophy, similarly used natural philosophy to elucidate his medical treatises, especially in his greatest work, *The Canon of Medicine*. When these works were translated into Latin as part of the Greco-Arabic corpus, the late Middle Ages inherited a long and rich tradition in which natural philosophy was applied to medicine.

Because of the structure of the universities, the links between natural philosophy and medicine were considerably strengthened during the late Middle Ages. At the great medical centers, such as those at the universities of Paris, Bologna, Montpellier, and Padua, many physicians, if not the majority, who received their degrees in these institutions had matriculated through a university arts program, perhaps even receiving a master of arts degree prior to entering as a student in the medical faculty. At Montpellier, a statute of 1240 required that entering medical students be competent in the arts. In Italy, medical students customarily taught the arts while they matriculated in the medical faculty. At the Universities of Bologna and Padua, arts and medicine were taught together within a single faculty. The average university-educated physician was thus trained in both arts and medicine and could, presumably, interrelate them more authoritatively than ever before.

Although university-trained physicians were in a minority among healers in the late Middle Ages, medical literature was produced overwhelmingly by physicians with university backgrounds. It was not unusual for medieval physicians to write on natural philosophy, as well as on medicine, especially in Italy, as Peter of Abano (1257–ca. 1316) did. Peter, who studied at the University of Paris and taught medicine and philosophy at the University of Padua, declared that logic, natural philosophy, and astrology were necessary adjuncts to the study of medicine: "Logic, since it is the condiment of all the sciences, just as salt is of food; natural philosophy, since it shows the principles of everything; and astrology since it is directive of judgments."[26] Nancy Siraisi has observed that "leading medical writers of the thirteenth to fifteenth centuries were ... the products of extensive training in logic and natural philosophy, and increasingly as time went on, in astrology as well."[27] Many university-trained physicians who wrote medical treatises were thus in a position to use their knowledge of natural philosophy to enhance their books. In so doing, they exhibit an intimate knowledge of Aristotle's natural books and thereby reveal the importance of natural philosophy for medicine. Indeed, medical authors frequently used the questions format to present their ideas, often basing their questions on the works of Galen and Avicenna. Occasionally, the questions were divorced from any text and simply covered a wide range of topics. Between 1290 and 1305,

Peter of Abano wrote his *Reconciler of the Differences among Philosophers and Especially among Physicians* (*Conciliator differentiarum philosophorum et praecipue medicorum*) in which he sought to reconcile medicine and natural philosophy. The questions were based on his medical lectures at the University of Paris, but they ranged over the whole of natural philosophy.

## Music

Of the quadrivial sciences of arithmetic, geometry, astronomy, and music, the first three were not penetrated by natural philosophy. To the contrary, it was natural philosophy that absorbed some mathematics and certain basic elements of astronomy. From an a priori standpoint, music would also seem an unlikely candidate for the utilization of natural philosophy. But in at least one instance, the anonymous *Questions on Music* (*Questiones musice*; perhaps by Blasius of Parma), a work on music, was composed in the scholastic mode in the late fourteenth century. The author included what one might expect in a music treatise, but, in his analysis of the questions, he applied numerous concepts developed in fourteenth-century natural philosophy. He found occasion to introduce the intension and remission of forms (applied to sound and to stringed instruments), the possibility of motion in a void, the perfection of species, relationships between infinites, first and last instants, and a number of other topics.

## CHARACTERISTIC FEATURES OF MEDIEVAL NATURAL PHILOSOPHY

With the exception of an occasional question, such as whether the earth is spherical, to which a precise affirmative answer was routinely supplied, we can now see that although the questions on Aristotle's natural books were answerable, they were not definitively answerable. Although Aristotle had shown how the syllogism was to be used to produce scientific demonstrations in natural philosophy, we have seen in this chapter that many medieval natural philosophers believed that questions and problems in natural philosophy could not be scientifically demonstrated by reason or experiment. Only probable, plausible, or conjectural responses could be provided for hundreds of questions, such as "whether the celestial spheres are moved by one or by several intelligences"; or "whether the heaven is moved with exertion and fatigue"; or "whether a comet is a terrestrial vapor"; or "whether the middle region of air is always cold"; or "whether elements remain formally in a compound [or mixed] body." Aristotelian natural philosophers were unable to formulate decisive answers for most such questions. They could only try to answer each question in accordance with the metaphysical and physical principles currently in vogue. The primary tools of analysis were metaphysics, theology, and counterfactuals, often in the form of thought experiments expressed in the language of natural impossibilities. Within such a context, experience and observation played small roles. Verification of truths in natural philosophy was rarely made by appeal to experiments or experience. Issues were decided by proper reasoning from a priori principles and truths. Given certain principles, things had to be this way, or that way, or yet another way. Only rarely could experience decide between possibilities. Because authors used the tools of analysis in quite different ways, disagreements were common. A given question might therefore receive a variety of responses from among the numerous authors who considered it. Most questions were in this category.

Were medieval natural philosophers disturbed by their seeming inability to come to widespread agreement on most of the questions they posed? If they were, there is no evidence of it. Perhaps they were not upset because, as was shown earlier in this chapter, they did not think of natural philosophy as an exact science. Most would have agreed with Blasius of Parma, who, in contrasting the derivation of mathematical conclusions from first principles, declared that "in natural philosophy this is not the case because the matter of natural things is not subject to demonstration."[28] We have already seen that agreement in natural philosophy was in the macrostructure and on certain fundamental principles, rather than in the operating details. Medieval natural philosophy was largely a matter of applying Aristotelian concepts to a host of problems within the overall macrostructure. Natural philosophers might agree, for example, that there are real orbs in the sky, but disagree as to precisely how many. As for the question about the number of intelligences that move a sphere, whether one or more, the answer would have been obvious, though indemonstrable, because of the Aristotelian metaphysical principle that for every orb there could be only one mover. This kind of precision, if it may be called that, was characteristic of medieval natural philosophy and seems to have satisfied the natural philosophers who produced it. We have no evidence that they were dissatisfied with multiple responses to a given question or that they believed that better answers had to be found.

Medieval natural philosophers were unlike early modern scientists in two significant ways: they did not regularly employ experiments as a means of gaining knowledge about the world, and they lacked a useful concept of scientific progress. Their failure to develop an experimental method may derive from the concept of substance, which they inherited from Aristotle. As Sarah Waterlow has explained it, for Aristotle every substance possessed an inner principle that guided its development and existence and that almost always operated harmoniously with its external environment. The idea of science is to identify and understand the workings of these inner principles, or forms. In a world of such substances, controlled experiments would do little good, because they would interfere with the regular environment of any given substance and thereby prevent us from learning about its true nature. Experiments that did not interfere with the environment of a substance would consequently provide no more information than could be obtained by observing its natural operation. Hence experiments would be at worst obstructive, and at best superfluous.[29] Although some experiments were carried out during the Middle Ages, they were largely about the reproduction of known effects, such as the rainbow or magnetism. The routine incorporation of experiments into science was a contribution of the Scientific Revolution.

Science today is almost synonymous with the advance of knowledge and with progress. Was this also true for the Middle Ages? Was natural philosophy associated with the advance of knowledge? Did natural philosophers believe that knowledge was cumulative and progressive? Did medieval natural philosophers have a concept of progress? Expressions of progress, especially technical progress, were not lacking in the Middle Ages. One of the most interesting was by the Dominican Brother Giordano of Pisa, who, in 1306, declared that

> Not all the arts have been found; we shall never see an end of finding them. Every day one could discover a new art.... indeed they are being found all the time. It is not twenty years since there was discovered the art of making spectacles which help you to see well, and which is one of the best and most necessary in the world. And that is such a short time ago that a new art, which never before existed, was invented.... I myself saw the man who discovered and practiced it, and I talked with him.[30]

There is, however, no plausible evidence as yet to indicate that such a concept played a role in natural philosophy. Natural philosophers did not perceive their role as one of promoting, or celebrating, the advancement of knowledge, although they occasionally suggested that they had advanced beyond the ancients. Indeed, as will be seen, they also recognized that Aristotle was wrong in certain instances, that his responses to problems were inadequate, and that he had not considered certain problems that confronted his commentators. They saw themselves as supplying correctives and additions and, therefore, in some sense, advancing beyond Aristotle. In the fourteenth century, natural philosophers coined terms such as "the modern way" (*via moderna*) and "the ancient way" (*via antiqua*) to distinguish those who followed the more recent philosophical views, designated loosely by the terms "nominalism" or "terminism," and associated with William of Ockham, and those who followed the older realist philosophies, associated with the names of

Thomas Aquinas and Duns Scotus. As a group, however, medieval natural philosophers were convinced that Aristotle's metaphysics and natural philosophy, along with their corrections and additions, were sufficient to determine all that could be known about nature. Perhaps they thought – if they thought about it at all – that they had only to apply the basic principles of Aristotelian natural philosophy to fill in the remaining gaps in their knowledge. This implies a sense of a continuing accumulation of knowledge until, in principle, everything worth knowing about the world would be known. It is unlikely that medieval natural philosophers believed that such a stage could ever be reached. Belief in the end of the world in the not-too-distant future, followed by a Judgment Day, would probably have precluded such an idea.

Among some, perhaps among many, natural philosophers in every generation there was probably a sense that they were the "moderns" and that they had advanced in some ways beyond their predecessors, and even beyond Aristotle himself. And yet little was made of this. In the Middle Ages, such feelings would have been professed, if they were professed at all, with humility, in the manner of Bernard of Chartres, who, in the twelfth century, observed that if we see farther than our predecessors, it is because we are "like dwarfs standing on the shoulders of giants."[31] But William of Conches argued that if a dwarf put on the shoulders of a giant sees farther than the giant, it does not follow that he is wiser than those who could not see as far. Transferring the dwarf-giant relationship to that between ancients and moderns, William explained that "The ancients had only the writings which they composed themselves, but we have all their writings and all those as well which were composed from the beginning up to our time. Hence we see more, but do not know more."[32] Such sentiments were not likely to translate into an idea of progress in natural philosophy, perhaps because, as we have seen, most questions and problems in natural philosophy lacked definitive answers. Advance and progress were thus difficult, if not impossible, to measure and define. Under such circumstances, natural philosophers would have had difficulty in arriving at a sense of progress, to say nothing of a sense of inevitable progress.

## ARISTOTELIANS AND ARISTOTELIANISM

Although there were no Latin terms in the Middle Ages for "Aristotelians" and "Aristotelianism," modern scholars use such terms because they are useful and descriptive. Because medieval natural philosophy was almost wholly based on Aristotle's natural books, it seems reasonable to call those who studied those works and wrote commentaries on them "Aristotelians." It is but a step from this to the term "Aristotelianism," which signifies the body of literature that was produced by Aristotelians and, simultaneously, represents an ill-defined set of attitudes and assumptions held by the majority of medieval natural philosophers about the structure and operation of the physical world. Although these descriptions may serve our immediate purposes, the problem of adequately defining the terms "Aristotelian" and "Aristotelianism" is difficult and paradoxical, largely because the relationship between medieval interpretations of Aristotle's ideas and what Aristotle may have really intended defies easy analysis and also because of an absence of reliable criteria to determine whether radical departures from Aristotelian ideas fall within the domain of Aristotelianism. Fortunately, although problems about departures, and alleged departures, from Aristotle are of interest, they represent a relatively minor problem during the Middle Ages, because Aristotle's natural philosophy was completely dominant in Western Europe between 1200 to 1450. It had no rivals. Real or imagined departures from the texts of Aristotle were not perceived as forming part of a "new," non-Aristotelian, world view, but were regarded as belonging to Aristotelianism. The authors of these ideas had to be regarded as Aristotelians. What else could they be? There were no other viable philosophies with which their divergent ideas might be identified.

Only in the sixteenth and seventeenth centuries was Aristotelianism first seriously challenged. Departures from Aristotelian natural philosophy became a significant factor with the introduction into Western Europe of new world views, a process that began in the second half of the fifteenth century, when the works of Plato were translated from Greek into Latin, and continued into the sixteenth century. Thus began another great wave of translations, this time almost exclusively from Greek into Latin. New doctrines and philosophies emerged as rivals to Aristotelianism. From Greek manuscripts preserved in European archives, or brought to Europe by Greeks fleeing the Turkish onslaught on Constantinople during the fifteenth century, came the doctrines of Stoicism, Platonism, Neoplatonism, Hermeticism, and atomism. At least one classical Latin treatise played a significant role in the new science that was developing. After centuries of obscurity, Lucretius's Latin treatise, *On the Nature of Things* (*De rerum natura*), emerged in the early fifteenth century to become a major cosmic system and the most complete report of atomism known. It served as a counterweight to Aristotle's hostile account, which up to that time had been the major exposition of atomism in Europe. Under pressure from these rival philosophies, traditional Aristotelianism was compelled to change and adapt. Therefore, the nature and definition of Aristotelianism also changed, as did the concept of an Aristotelian. "Aristotelians" and "natural philosophers" were no longer coextensive terms. Distinguishing an Aristotelian natural philosopher from a non-Aristotelian natural philosopher is not always easy.

It has sometimes been argued that the essence of Aristotelianism lies in a hard core of basic, general principles to which all medieval natural philosophers subscribed, and which none challenged. That is what makes them Aristotelians. Disagreements arise only in the application of these principles to a host of problems and situations in the physical world. What are these fundamental principles and truths? Foremost among them are that there is a radical dichotomy in the world between the celestial and terrestrial regions, whereby the former is distinguished by its incorruptible celestial ether and the latter by its corruptible matter; that there are four terrestrial elements; that there are four causes that operate in the world; that there are four primary qualities; that there is an ultimate cause of all motion, the Prime Mover; and that the earth lies immobile in the geometric center of the universe.

Unfortunately, it is difficult to determine what the medieval natural philosophers regarded as the core of Aristotelian principles. For example, one of Aristotle's foundational principles was his assumption that the world is eternal, without beginning or end. But we saw that this assumption was condemned at Paris in more than twenty-five versions. Although scholastics could utilize the eternity of the world for argumentative purposes, no one upheld it explicitly during the Middle Ages. Aristotle's principle that celestial and terrestrial matter are radically different was also challenged during the Middle Ages by Giles of Rome and William of Ockham, and others. The earth's location at the center of the universe was also dubious, because Ptolemaic astronomy required that the earth be eccentric and therefore removed from the world's geometric center, even as Aristotelian natural philosophers routinely spoke of its centrality.

From these few instances, we can obtain a sense of the obstacles to defining concepts like Aristotelian and Aristotelianism. During the Middle Ages, however, when Aristotelianism had no rivals and all natural philosophers were Aristotelians by default, the problem of defining the terms was of no consequence. But in the sixteenth and seventeenth centuries the problems increased. Aristotelians were no longer a homogeneous group. Indeed, a few self-proclaimed Aristotelian natural philosophers even adopted the Copernican heliocentric system (Thomas White [1593–1676]) or elements of it such as the axial rotation of the earth (Andreas Cesalpino [1519–1603]). Other seventeenth-century scholastic natural philosophers accepted a corruptible heaven, thereby abandoning Aristotle's strict dichotomy between the celestial and terrestrial regions.

Despite having no real rivals during the Middle Ages, Aristotelianism was not a rigid body of doctrine slavishly defended by its adherents. Although Aristotle was greatly admired (see the tributes to him in chapter 4), Albertus Magnus offered an interesting insight into the way many medieval natural philosophers probably approached Aristotle. Aristotelians, or "peripatetics," as Albertus called them, agreed that "Aristotle spoke the truth, for they say that Nature set up this man as if he were a rule of truth in which she demonstrated the highest development of the human intellect – but they expound this man in diverse ways, as suits the intention of each one of them."[33] Although Albertus admired Aristotle and believed that he intended to speak the truth, he did not regard him as infallible. The attribution of infallibility would be warranted only if Aristotle were a god, but "if, however, one believes him to be but a man, then without doubt he could err just as we can too."[34]

On many points there was disagreement about Aristotle's "real" meaning. Attempts to interpret Aristotle's intent during the Middle Ages were made difficult because his works were treated as if he had written all of them simultaneously. Indeed, they were treated like extra-temporal productions, without chronology or context, as if Aristotle's thought had never developed over time, but had sprung from his brain full-blown. Instead of interpreting Aristotle's conflicting opinions on the same theme as the consequence of a possible change of mind in a later work, medieval scholars believed that they had to reconcile such opinions arbitrarily. It would have been thought unseemly to accuse Aristotle of self-contradiction or of a change of mind. As a revered authority, Aristotle's opinions were not to be set aside lightly, except where they conflicted with the Christian faith or where he was perceived to be manifestly wrong. In such instances, attempts were made at reconciliation or correction.

Although scholastic authors were usually reluctant to draw attention to inadequate or mistaken interpretations in Aristotle's thought, they seem to have done so on some rather important issues. Sometimes, in their acknowledged departures from Aristotle, authors would seek to "save" him from the embarrassment of error. Thomas Aquinas, for example, did this in his *Commentary on Aristotle's Physics*, where he rejected Aristotle's argument that a motion in a vacuum could bear no ratio to a motion in a plenum. Aquinas argued that Aristotle did not really seek a "demonstrative" argument and added, a few paragraphs later, that in this fourth book of his *Physics* Aristotle chose to introduce some false notions because he was considering the specific and definite natures of bodies rather than the nature of body in general.

Although Nicole Oresme also proposed a face-saving alternative in behalf of Aristotle, he allowed for the possibility that Aristotle had failed to understand the problem. Oresme explicitly rejected as false two rules of motion that Aristotle expressed in the seventh book of his *Physics*. In Oresme's words, Aristotle held that "if a power moves a mobile with a certain velocity, double the power will move the same mobile twice as quickly" (that is, if $F/R \propto V$, then $2F/R \propto 2V$) and "if a power moves a mobile, the same power can move half the mobile twice as quickly" (that is, if $F/R \propto V$, then $F/(R:2) \propto 2V$) (for Aristotle's rules, see chapter 4). To

replace Aristotle's false rules expressed in terms of arithmetic proportionality, Oresme presented his own version, based on geometric proportionality (see chapter 6). "What, then," said Oresme, "should we say
to Aristotle who seems to enunciate the repudiated rules in the seventh
[book] of the *Physics*?" At first, Oresme attempted to "gloss" Aristotle's
text by rephrasing the rules so that they are "correct"; he then suggested
that perhaps Aristotle really had the right version, but it was poorly
translated. Finally, however, he allowed for the possibility that Aristotle
did not understand the rules properly. Thus Oresme first sought to save
Aristotle by admitting that he might have had the right formulation,
which somehow became defective in translation. Only thereafter did he
suggest that Aristotle may have failed to understand the rules and fallen
into error.

On at least two occasions, John Buridan was directly critical of Aristotle, without offering alternative face-saving explanations. The first
occurs in Buridan's *Questions on De caelo* (bk. 1, qu. 18), and involves
Aristotle's discussion of the possibility of the existence of other worlds,
where Aristotle argued that the earth of this world would be moved not
only to the center of its own world but also to the center of another
world. Buridan rejected Aristotle's explanation and characterized it as
"undemonstrative," arguing that our earth would not move to the center
of another world, but would remain at rest in the center of our world.[35]
The second occasion appears in Buridan's *Questions on the Physics*, where
he asked "whether a projectile after leaving the hand of the projector is
moved by the air, or by what is moved" and declared that "the question"
is "very difficult because Aristotle, as it seems to me, has not solved it
well."[36] Buridan rejected Aristotle's claim that the external air moves a
body after it has left the hand of the projector. As a replacement for
Aristotle's ill-conceived explanation, Buridan proposed his famous impetus, or impressed force, theory to account for the continuing motion
of projectiles that have lost contact with their initial projectors.

Finally, there were natural philosophers like the theologian-natural
philosopher Nicholas of Autrecourt (see earlier, this chapter), who were
convinced that properly demonstrated truths were unattainable and
sought to abandon Aristotle's conclusions and replace them with, at
least, equally probable alternatives based on atomism, which Aristotle
had vigorously rejected. Very few trod this path, which represents an
unusual attempt to reject Aristotle's natural philosophy.

Medieval natural philosophers interpreted Aristotle in a variety of
ways, sometimes taking issue with him. Because he was a highly regarded authority, scholastic natural philosophers, as we have seen, were
often willing to give Aristotle the benefit of any doubts they may have
entertained about his intentions and meanings. But their respect for Aristotle, which often bordered on reverence, did not prevent them from

taking numerous significant departures from his plain intent. These departures ranged over most of Aristotle's natural philosophy. For convenience, two kinds may be distinguished. The first were substantive, either based upon perceived discrepancies, inconsistencies, and weaknesses in Aristotle's system or expansions of ideas that Aristotle may have merely suggested, or left undeveloped. The second kind were hypothetical assumptions about the physical world that were impossible by natural means in Aristotle's philosophy. These were attributed to God's infinite and absolute power.

Among the first kind of departure, we may include the assimilation of eccentric and epicyclic orbs into Aristotle's system of concentric spheres and the manner in which they affected Aristotelian cosmology and physics; the relationship between celestial and terrestrial matter; the change from motive causes that were external, such as air, to internal, or impressed, forces; and the ways in which qualities vary, or, to use the medieval expression, the intension and remission of forms or qualities. Of the hypothetical assumptions in the second category, some were derived from articles condemned in 1277, especially those concerned with what might lie beyond our world, namely, the possibility of other worlds and the possibility of infinite, void space. Because these departures form one of the most interesting aspects of medieval natural philosophy, my discussion in chapter 6 focused on them.

Despite these significant departures from Aristotelian natural philosophy, Aristotelianism did not, and probably could not, transform itself into something new. From its entry into Europe, Aristotelian natural philosophy was extraordinarily broad. Over the centuries, much that was fundamental in Aristotelianism was challenged. Aristotle's medieval followers frequently altered their opinions, replacing earlier interpretations with new ones. The new and the old were sometimes retained, so that Aristotelianism included conflicting opinions. It was thus a vast body of natural knowledge that was both elastic and absorbent. As the centuries passed and rival philosophies entered Europe, these tendencies intensified. By the sixteenth and seventeenth centuries, some Aristotelian natural philosophers sought to accommodate to the new views, especially to the new science that had been developing since the advent of Copernicanism. It was too late. The effort was unsuccessful. Aristotelianism could not be transmuted into anything resembling Newtonianism. It had ceased to be a respectable rival of the new science. Whatever Aristotelian natural philosophy still possessed that was useful had to be incorporated into the new science, into a quite different way of looking at the world.

Why was Aristotelian natural philosophy never reformed from within? Was it even capable of such a reformation? Probably not. Aristotelianism was too all-encompassing and unwieldy. If medieval natural philosophers had been more self-critical about their broadly conceived discipline, they might have recognized that "Aristotle's was the most capacious of philosophies," because "in principle it explained everything," as Charles Gillispie has perceptively characterized it.[37] Aristotelian physical principles, such as potentiality-actuality, the four causes, matter and form, the four elements, the doctrine of natural place, were so broad and comprehensive that they were easily applied to rival theories and arguments. Not only were these basic principles rarely challenged explicitly, but they found a range of application that would have surprised, if not shocked, Aristotle himself.

Most of the opinions and interpretations in Aristotelian natural philosophy were essentially irrefutable, giving them an air of indestructibility. As a natural philosophy in which mathematics, experiment, and prediction played insignificant roles in discovering the structure and operations of nature, no counterarguments or proofs could easily affect it, to say nothing of demolishing it. Individual Aristotelians simply adopted some of the new ideas and proceeded to incorporate them into the available matrix of Aristotelianism, which simply grew larger and more complex, if not incoherent. No serious effort was made to mesh the new with the old to forge a more viable Aristotelianism. In this, Aristotelianism differed from two other scientific "isms" – Newtonianism and Darwinism – whose supporters strove mightily to improve the consistency of their respective systems. When mathematical physics reached a stage of development in the seventeenth century where it could provide explanations for and predictions of phenomena that were impossible within Aristotelian natural philosophy, the latter's fate was sealed. Aristotelianism did not evolve into anything else. It simply faded away.

In truth, confidence in Aristotelian natural philosophy had been steadily eroding during the sixteenth and early seventeenth centuries. Two momentous events more than a century apart subverted it. The discovery of the New World by Columbus in 1492 destroyed the ancient and medieval view of the earth. It made European scholars realize that Aristotle's knowledge of the earth, and that of the ancients generally, was not only greatly limited but also often flatly wrong. It began the lengthy process that would end with the destruction of Aristotelianism. The second prodigious event occurred in 1610 when Galileo turned the recently invented telescope skyward and discovered the satellites of Jupiter, as well as stars that had never before been seen. Galileo did to the cosmos what Columbus had done to the earth: he revealed the inadequacy of ancient and medieval knowledge. A new world would be fashioned with new knowledge derived in ways that were previously unknown. Many came to believe that Aristotelian natural philosophy had failed to reveal, and was perhaps incapable of revealing, nature's previously hidden phenomena, its myriad of "secrets."

Enough has now been said about medieval natural philosophy as a subject and discipline. We must now see what role it played in laying the foundations of modern science.

# 8

# How the foundations of early modern science were laid in the Middle Ages

ALTHOUGH science has a long history with roots in ancient Egypt and Mesopotamia, it is indisputable that modern science emerged in the seventeenth century in Western Europe and nowhere else. The reasons for this momentous occurrence must, therefore, be sought in some unique set of circumstances that differentiate Western society from other contemporary and earlier civilizations. The establishment of science as a basic enterprise within a society depends on more than expertise in technical scientific subjects, experiments, and disciplined observations. After all, science can be found in many early societies. In Islam, until approximately 1500, mathematics, astronomy, geometric optics, and medicine were more highly developed than in the West. Indeed, the West learned these subjects from translations of Arabic treatises into Latin. But science was not institutionalized in Islamic society. Nor was it institutionalized in ancient and medieval China, despite significant achievements. Similar arguments apply to all other societies and civilizations. Science can be found in many of them, but it was institutionalized and perpetuated in none.

An impressive array of scholarship testifies that modern science emerged in Western Europe as a result of the Scientific Revolution, a phenomenon associated overwhelmingly with the seventeenth century. That same scholarship has proclaimed that the emergence of modern science in the seventeenth century owes little or nothing to the Middle Ages. Not only did medieval natural philosophy play little, if any, role in the advent of early modern science, so goes the argument, but it was the major obstacle to it. After all, was it not Aristotelian medieval natural philosophy that Galileo mercilessly attacked when he created the character of Simplicio (in his *Dialogue on the Two Chief World Systems*), the stereotypical, dull-witted, unimaginative, unyielding scholastic defender of the indefensible? In Galileo's mind, scholastic natural philosophy was the enemy of the new science that he and others were trying to forge. Support for Galileo became commonplace as the seventeenth century progressed. By the end of that century, the new science was triumphant and seemed to owe nothing to approximately five centuries of medieval natural philosophy. By then, Aristotelianism was in disarray. As time passed, scholastic literature on natural philosophy was read less and less, and, by the twentieth century, the treatises and their authors had virtually disappeared from scholarly discourse.

Despite a heroic effort in the twentieth century to revise this negative judgment about medieval science and natural philosophy, it has had little impact. The opinion of educated people today probably differs little from Galileo's assessment, which was adopted wholeheartedly by the science establishment and is now deeply embedded in our culture. After some three centuries, the overwhelming consensus is that modern science began with the likes of Copernicus, Galileo, Kepler, Descartes, and Newton and has no connection with the preceding centuries of scholastic natural philosophy. Indeed, it is widely believed that the new science triumphed in spite of the obstacles posed by reactionary Aristotelians.

Yet this interpretation is misleading, largely because it is woefully incomplete. Its very incompleteness constitutes a powerful obstacle to a better understanding of the development of early modern science. The assault on scholasticism by the new breed of natural philosophers and scientists in the sixteenth and seventeenth centuries was quite appropriate. By the sixteenth century, traditional scholasticism needed criticism, even though scholastic natural philosophers had made noteworthy contributions to physical thought, a number of which have been mentioned in this volume. Their contributions on various aspects of motion, both kinematic and dynamic, were striking; Galileo himself could not actually enhance them. Sophisticated discussions about the earth's possible axial rotation were also impressive, and Copernicus could not improve upon them. Medieval ideas about other worlds, and especially about infinite void space, played an unacknowledged role in shaping the cosmos that was being constructed by the new science.

Other alleged medieval contributions to the new science are in dispute. At best, one can show parallels but rarely direct influence. Did Galileo derive the mean speed theorem and the various definitions of speed and acceleration from printed medieval texts available in the sixteenth century? Did Copernicus derive some of his arguments about the earth's daily rotation from scholastic texts, perhaps those of John Buridan, at the University of Cracow in the late fifteenth century? Were Henry More and Isaac Newton the indirect beneficiaries of medieval ideas about the intimate relationship between God and space?

What about medieval contributions to natural philosophy that came from outside the universities? William Newman's recent research on medieval and early modern alchemy seems to reveal that seventeenth-century corpuscular theory had deep medieval roots.[1] Other possible continuities might also be found.

Even if linkages could be shown in each of these cases, and in others as well, it is conceivable that a majority of scholars might conclude that the cumulative impact of these influences still did not warrant the inference that medieval natural philosophy significantly affected the content and direction of the physical science that was being shaped in the seventeenth century. (At the very least, however, those linkages would serve to reinforce the major claim of this chapter, and ultimately of this study, namely, that there was continuity between medieval and early modern natural philosophy.) For even if Galileo had benefited from medieval discussions about the mean speed theorem, so the argument might go, it is evident that he transformed those hypothetical arguments into the science of mechanics and made them the beginning of modern physics. If Newton profited from medieval arguments about God and infinite space, he went beyond anything envisioned in those arguments when he made that infinite space the frame for the motion of bodies that he assumed to move everywhere in that space. Whether similar arguments could be made for other claims about possible continuities between medieval and early modern science and natural philosophy remains for future research to determine.

Claims about medieval influence on seventeenth-century science will continue to be controversial. But whether some, or all, of them are ultimately accepted or rejected is irrelevant for the thesis of this study. The links between medieval and early modern science that I propose are independent of specific claims about this or that influence, in this or that science.

If we cannot readily and unambiguously detect significant medieval scientific influences on seventeenth-century science, might it not be reasonable to assume that there were none? Perhaps the traditional consensus that denied any role to the Middle Ages in the birth of the Scientific Revolution was appropriate after all. To those who might make such a claim, I respond by appeal to the argument in the preface of this book: a scientific revolution could not have occurred in Western Europe in the seventeenth century if the level of science and natural philosophy had remained what it was in the first half of the twelfth century, that is, just prior to the translation of Greco-Arabic science that was under way in the latter half of that century. Without the translations, which transformed European intellectual life, and the momentous events that followed from them, the Scientific Revolution in the seventeenth century would have been impossible.

It is in the Latin Middle Ages in Western Europe that we must look for answers to questions such as: Why did science as we know it today materialize only in Western society? What made it possible for science to acquire prestige and influence and to become a powerful force in Western Europe by the seventeenth century? The answers to such questions are to be found in certain attitudes and institutions that were generated in Western society from approximately 1175 to 1500. These attitudes and institutions were directed toward learning as a whole and toward science and natural philosophy in particular. Together they coalesce into what may be appropriately called "the foundations of modern science." They were new to Europe and unique to the world. Because there is nothing to which we can compare this extraordinary process, no one can say whether it was fast or slow.

In the first seven chapters of this book, I described the new factors that transformed Western Europe from an intellectual embarrassment to an intellectual powerhouse. I shall now use two distinct approaches to show why, collectively, they formed the foundations of modern science. The first involves considering the contextual changes that created an atmosphere conducive to the establishment of science. Here I shall be concerned with the conditions that made it feasible to pursue science and natural philosophy on a permanent basis and that also made those pursuits laudable activities within Western society. In the second approach I shall examine certain features of medieval science and natural philosophy that were favorable to the development of the Scientific Revolution.

## THE CONTEXTUAL PRE-CONDITIONS THAT MADE THE SCIENTIFIC REVOLUTION POSSIBLE

The creation of a societal environment in the Middle Ages that eventually enabled a scientific revolution to develop in the seventeenth century involved at least three crucial pre-conditions: (1) the translation of Greco-Arabic works on science and natural philosophy into Latin, (2) the formation of the medieval university, and (3) the emergence of the theologian-natural philosophers.

### The translations

The first, indispensable, pre-condition for the Scientific Revolution was the translation of Greco-Arabic science and natural philosophy into Latin during the twelfth and thirteenth centuries. Because of the importance of the translated works, the civilization of Islam must be allotted a considerable share of the glory for the Western achievement in science. Centuries before, Islamic scholars had translated a large part of Greek science into Arabic and then had added much that was original, to form what is conveniently referred to as Greco-Arabic (or Greco-Islamic) science, at the core of which lay the works of, as well as the commentaries on, Aristotle. This large body of learning was subsequently transferred to the Western world. Although science in the West might have developed without benefit of this Greco-Arabic legacy, the advent of modern science would undoubtedly have been delayed by centuries, if not yet still lie in the future.

## The universities

A second pre-condition for the Scientific Revolution was the formation of the medieval university, with its corporate structure and varied activities. The universities that had emerged by the thirteenth century in Paris, Oxford, and Bologna were different from anything the world had ever seen. Nothing in Islam, China, or India, or in the ancient civilizations of South America, was comparable to the medieval university. It is in this remarkable institution, and its unusual activities, that the foundations of modern science must be sought.

The university as we have come to know it, and as it was described in chapter 3, was made possible because the evolution of medieval Latin society permitted the separate existence of church and state, each of which was willing to recognize the separate existence of corporate entities such as the university. A connection between the development of the medieval university and translation of Greco-Arabic works into Latin is probable. The Universities of Paris, Oxford, and Bologna were established by approximately 1200, shortly after most of the translations had been completed. The translations furnished the emerging universities with a ready-made curriculum, one comprised primarily of the exact sciences, logic, and natural philosophy. If those translations had been primarily of treatises in literature, poetry, and history, the university curriculum would have been radically different. The translations, however, were overwhelmingly in the realm of science and natural philosophy. The incorporation of numerous translated works, especially the works of Aristotle, as well as of original works based on the translations, into the curriculum of the universities allowed for the institutionalization of science and natural philosophy.

The curriculum of science and natural philosophy established in the medieval universities of Western Europe was a permanent fixture for approximately four hundred fifty to five hundred years. Courses in logic, natural philosophy, geometry, arithmetic, music, and astronomy constituted the subjects of study for the baccalaureate and master of arts degrees. These disciplines formed the basis of the curriculum of the arts faculty, which, as we saw in chapter 3, was the largest of the traditional four faculties – the others being medicine, theology, and law – of a typical major university.

For the first time in history, an institution had been created for the teaching of science, natural philosophy, and logic. Also for the first time, an extensive four- to six-year course of study in higher education was based on a fundamentally scientific curriculum, with natural philosophy as its most important component. Even more remarkable is that these disciplines served as the core curriculum for all students and were virtual prerequisites for entry into the higher disciplines of law, medicine, and theology. They were taught on a regular basis for centuries. As universities multiplied during the thirteenth to fifteenth centuries, the same logic-science-natural philosophy curriculum was disseminated throughout Europe, extending as far east as Poland.

The introduction of Aristotelian science and natural philosophy into Western Europe, from Greek and Arabic sources, furnished the basis of a curriculum for the emerging universities. Without the tacit approval of church and state, however, that curriculum could not have been implemented. To a remarkable extent, church and state granted to the universities corporate powers to regulate themselves, thus enabling the universities to determine their own curricula, to establish criteria for the degrees of their students, and to determine the teaching fitness of their faculty members.

Despite some disciplinary tensions and difficulties that arose in the thirteenth century between the arts faculty, representing natural philosophy and reason, and the faculty of theology, representing theology and revelation, arts masters and theologians alike welcomed the arrival of Aristotle's natural philosophy. They revealed their approval by making it the foundation of the curriculum for higher education. Why did they do this? Why did a Christian society at the height of the Catholic Church's power readily adopt a pagan natural philosophy as the basis of an extensive educational program? Why did Christians enthusiastically embrace, rather than reject, such a program?

They embraced it because, as we saw in chapter 1, Christians had long before come to terms with pagan thought and were agreed, for the most part, that they really had little to fear from it. The "handmaiden" attitude that initially defined Christianity's attitude toward pagan thought had made acceptance of Greek natural philosophy tenable. When Christians in Western Europe learned of the existence of Greco-Arabic scientific literature and were finally ready to receive it, they did so eagerly. Although religious authorities at Paris had some difficulty with Aristotle's thought, even going so far as to condemn a number of ideas drawn from his books, the works of Aristotle were challenged only for relatively short periods, confined largely to the thirteenth century. The Condemnation of 1277 was a local effort in which the pope acquiesced after the fact, that is, after the condemnation had already been issued by the bishop of Paris. The Church itself, however, did not see fit to oppose the new literature. On the contrary, the introduction of Greco-Arabic natural philosophy and science in the twelfth and thirteenth centuries was welcomed as vital to intellectual life. Aristotle's books were made the basis of the university curriculum by 1255 in Paris, and long before that at

Oxford. Churchmen were eager to familiarize themselves with the works of Aristotle, as a striking statement from Albertus Magnus bears witness. Shortly before 1250, in Cologne, Albertus began his commentary on Aristotle's *Physics*, the first of a lengthy series of Aristotelian commentaries he wrote for his fellow Dominicans. At the outset, he proclaimed his purpose:

> Our intention in the science of nature is to satisfy (in accord with our capacity) brothers of our order, begging us for the past several years now that we might compose a book on physics for them of such a sort that in it they would have a complete science of nature and that from it they might be able to understand in a competent way the books of Aristotle. Although we consider ourselves insufficient for this task, nevertheless, since the requests of our brothers would not cease, finally we have undertaken what we had often refused, vanquished by the pleas of certain ones among them.[2]

As further testimony to the attachment of Western Europe to Aristotle's thought, I need only mention that manuscripts of his works, and the works of his commentators, were multiplied and disseminated in large numbers and, with the advent of printing, were printed often and widely.

What of the oft-mentioned charge that the Aristotelianism of the universities had lost touch with the changing times and was hostile to the new science emerging in the sixteenth and seventeenth centuries, thus forcing the latter to develop outside of the universities, especially in the newly established scientific societies? Although this may be an exaggeration, its truth or falsity does not affect the claims made here. Whatever the state of the universities in the seventeenth century, these venerable institutions had already done their foundational work. They had shaped the intellectual life of Western Europe. Their influence was everywhere.

### The theologian-natural philosophers

The third, and final, pre-condition for the Scientific Revolution was the emergence of a class of theologian-natural philosophers, that is, a class of individuals who were not only trained in theology – most had theological degrees – but who also had previously attained the degree of master of arts or its equivalent and were therefore thoroughly trained in that discipline. Their importance cannot be overestimated. If theologians at the universities had decided to oppose Aristotelian learning as dangerous to the faith, it could not have become the focus of study in European universities. Without the approval and sanction of these scholars, Greco-Arabic science and Aristotelian natural philosophy could not have become the official curriculum of the universities.

The development within the universities of Western Europe of a class of theologian-natural philosophers was extraordinary. Not only did they endorse a secular arts curriculum, but most believed that natural philosophy was essential for a proper elucidation of theology. Schools of theology expected their entering students to have attained a high level of competence in natural philosophy. As evidence of this, students who wished to matriculate for a theology degree were usually required to have acquired a master of arts degree. Because of the intimate relationship between theology and natural philosophy during the Middle Ages and because arts masters had been forbidden by oath (since 1272) to treat theological problems, it fell to the theologians to apply natural philosophy to theology and theology to natural philosophy. Their training in both disciplines enabled them to do so with relative ease and confidence, whether this involved, for example, the application of science and natural philosophy to scriptural exegesis, the application of the concept of God's absolute power to hypothetical possibilities in the natural world, or the invocation of scriptural texts to support or oppose scientific ideas and theories. Theologians had a remarkable degree of intellectual freedom to cope with such problems and rarely allowed theology to hinder their inquiries into the physical world. If there was any temptation to produce a "Christian science," medieval theologians successfully resisted it. Biblical texts were not employed to "demonstrate" scientific truths by appeal to divine authority.

The relatively small degree of trauma that accompanied Greco-Arabic science and natural philosophy into Western Europe and the subsequent high status that science and natural philosophy achieved in Western thought were in no small measure attributable to the class of theologian-natural philosophers. Despite having some anxiety about natural philosophy during the thirteenth century, theologians came to embrace it. They were significant contributors to both natural philosophy and science, and most of them lectured and wrote on these subjects. Indeed, some of the most noteworthy accomplishments in science and mathematics during the Middle Ages came from theologians, as the names of Albertus Magnus, Robert Grosseteste, John Pecham, Theodoric of Freiberg, Thomas Bradwardine, Nicole Oresme, and Henry of Langenstein bear witness. So enthusiastically did theologians embrace natural philosophy in their theological treatises that the Church had at times to admonish them to refrain from frivolously using natural philosophy to resolve theological problems.

By the late Middle Ages, Western Christianity had a long-standing tradition of using pagan thought for its own benefit. As supporters of that tradition, medieval theologians treated the new Greco-Arabic learning as a welcome addition that would enhance their understanding of Scripture. The occasional reactions against natural philosophy – as in the early thirteenth century, when Aristotle's works were banned for some years at Paris, and in the later thirteenth century when the bishop of Paris issued the Condemnation of 1277 – ought to be interpreted as relatively minor aberrations when viewed against the grand sweep of the history of Western Christianity.

The three pre-conditions just discussed – the translations, the universities, and the theologian-natural philosophers – laid a foundation for the emergence of modern science because they provided an environment that was conducive to the study of science. If no translations of Greco-Arabic science and natural philosophy had occurred in the twelfth and thirteenth centuries, and Europeans had been compelled to elevate their intellectual discourse without external assistance, it is inconceivable that a scientific revolution could have occurred in the seventeenth century. Nor could such a revolution have come to pass were it not for the long-standing existence of the science-natural philosophy curriculum of the medieval university. Many of the major problems that were confronted, and often resolved, in the sixteenth and seventeenth centuries were inherited from the Middle Ages. Without these problems, and without the long tradition of natural philosophy in the medieval universities, there would have been little to discuss in the seventeenth century. And without the support of the theologians and the Church, the medieval universities would have been unable to institute the science-logic-natural philosophy curriculum that began Western Europe's long, uninterrupted involvement with scientific thought and problems. Finally, to appreciate the importance of the theologian-natural philosophers to the emergence of modern science, it is instructive to compare the medieval European experience in relating natural philosophy and religion with the way in which these two disciplines were related in the civilizations of Islam and Byzantium.

*Religion and natural philosophy in medieval Islam.* In the Islamic world, for many centuries – say, from the ninth to the end of the fifteenth century – the level of science, especially in the exact sciences and medicine, far exceeded the level in Western Europe. By the close of the fifteenth century, however, Islamic science seems to have lost its momentum and entered a state of decline, just as science in the West began an upward surge that would produce major changes in the sixteenth century and, in the seventeenth century, culminate in a new physics, astronomy, and cosmology. The numerous hypotheses explaining why medieval Islamic science entered a period of stagnation cannot be described here. However, because a major thesis of this volume is that medieval natural philosophy in Western Europe played a significant role in producing the Scientific Revolution, I shall focus on the fate of that broad discipline rather than on the exact sciences (astronomy, optics, mathematics) and medicine.

The relationships that developed between natural philosophy and theology in the civilizations of Islam and Western Christianity were strikingly different. In Islam there were two kinds of science: the Islamic sciences, which were based on the Koran and Islamic law and traditions, and the "foreign sciences," or pre-Islamic sciences, which embraced ancient Greek science and natural philosophy. The foreign sciences that were translated into Arabic in the ninth and tenth centuries were eagerly received. So desirous were the Muslims for this knowledge that A. I. Sabra, the eminent historian of Islamic science, has argued that the process might better be characterized as an active "appropriation" rather than a passive "reception."[3] Over some four or five centuries, many scholars in Islamic society, including those who were Christians and Jews, absorbed Greek science and natural philosophy and added much to them. Nevertheless, the theologians and religious leaders who were the guardians of orthodoxy in Islamic society did not warmly receive these disciplines.

Natural philosophy was studied and interpreted by philosophers (*falāsifa*). Their approach to the world was largely based on the writings of Aristotle and on some Neoplatonic works that were attributed to Aristotle. In Islam, Greek natural philosophy was a mixture appropriately characterized as Neoplatonic Aristotelianism. Over the centuries, the number of identifiable Islamic philosophers is relatively small. Among these, the most eminent were al-Kindi, al-Farabi, Avicenna (Ibn Sina), Avempace (Ibn Bajja), and Averroes (Ibn Rushd). Some see Averroes, who died in 1198, as the last significant Islamic commentator on Aristotle.

Greek natural philosophy played a broader role in Islam when it was interwoven with the science known as *kalam*, which has been defined by A. I. Sabra as "an inquiry into God, and into the world as God's creation, and into man as the special creature placed by God in the world under obligation to his creator."[4] The *mutakallimun*, as the practitioners of *kalam* were collectively called, sought to support revealed truth and to reconcile it with rationality. To do so, they utilized Greek philosophy and its forms of argumentation, thus broadening its impact. Although Muslim theology, as practiced in *kalam*, involved considerable Greek philosophy, its study in *kalam* was not for its own sake. It was always the handmaiden of religion. Greek philosophy was used to defend and explain the Koran and its doctrines. In fact, many *mutakallimun* used their knowledge of Greek philosophy to attack it.

If we regard the *mutakallimun* as theologians, we may say that at least some Muslim theologians used Greek philosophy, whatever their motives may have been. But the *mutakallimun* probably represented only a small percentage of those whom we would want to call theologians. The great majority of Islamic theologians did little or no philosophizing and probably knew little about Greek philosophy. If they were aware of it, they probably rejected its use in connection with the Koran. Most Islamic theologians would have considered even the suggestion that Greek philosophy was needed to defend Islam and the Koran as blasphemous. Their interests were largely confined to the Koran, to Islamic law, and to the traditions of Islam.

Most Muslim theologians were convinced that on certain crucial issues Greek logic and natural philosophy – especially Aristotle's natural philosophy – were incompatible with the Koran. The greatest issue that divided the Koran from these disciplines was the creation of the world, which is upheld in the Koran, but was denied by Aristotle, for whom the eternity of the world was an essential truth of natural philosophy. Another serious disagreement arose in connection with the concept of secondary causation. In Islamic thought, the term "philosopher" was often reserved for those who assumed with Aristotle that natural things were capable of causing effects, as when a magnet attracts iron and causes it to move, or when a horse pulls a wagon and is seen as the direct cause of the wagon's motion. On this approach, God was not viewed as the immediate cause of every effect. Philosophers believed with Aristotle that natural objects could cause effects in other natural objects because things had natures that enabled them to act on other things and to be acted upon. By contrast, most Muslim theologians believed, on the basis of the Koran, that God caused everything directly and immediately and that natural things were incapable of acting directly on other natural things. Although secondary causation is usually assumed in scientific research, most Muslim theologians opposed it, fearing that the study of Greek philosophy and science would make their students hostile to religion.

Because of this perceived incompatability, Greek natural philosophy was usually regarded with suspicion and was therefore rarely taught publicly. Philosophy and natural philosophy were often outcast subjects within Muslim thought. Many of the foremost Muslim scientists and natural philosophers, including al-Biruni, Avicenna (Ibn Sina), and Al-hazen (Ibn al-Haytham), were supported by royal patronage and did not teach in schools. Without the protection of a strong patron, natural philosophers, who took Aristotle as their guide and mentor, were subject to attacks and denunciations by influential local religious leaders, who may have been offended by the doctrine of the eternity of the world and by the emphasis on reason that were characteristic of Aristotelian natural philosophy. In this regard, a description by F. E. Peters, who has studied the fate of Aristotle in Islam, is appropriate: "Assimilation there was –

the entire history of *kalam* shows it – but when confronted with radical *falsafah*, Islamic orthodoxy reacted with determination and frequently with violence."[5] Philosophy and natural philosophy were not to be studied for their own sakes because they might prove harmful to the faith (logic was often characterized as "ungodly"). Only to the extent that they served religion were they to be studied. Thus science and natural philosophy were subordinated to religion and were usually perceived as "handmaidens to theology." Arithmetic and astronomy were acceptable, for example, because they were regarded as indispensable to the faith, the former for dividing inheritances and the latter for obtaining values for astronomical phenomena essential for determining the times of the five daily prayers. It is ironic that in Islam, the handmaiden idea of the sciences took hold and remained. In the Latin West, by contrast, the handmaiden idea, which was dominant by the mid-fourth century, long before the introduction of Aristotelian natural philosophy in the twelfth century, faded away, as philosophy gradually became an independent discipline and as Aristotelian natural philosophy came to be valued for its own sake.

One of the greatest and most brilliant religious and philosophical writers in the history of Islam, al-Ghazali (1058–1111), was deeply suspicious and fearful of the effects on the Islamic religion of natural philosophy, theology (actually metaphysics), logic, and mathematics. In his famous quasi-autobiographical treatise, *Deliverance from Error*, he explained that religion does not require the rejection of natural philosophy, but that there are serious objections to it because nature is completely subject to God, and no part of it can act from its own essence. The implication is obvious: Aristotelian natural philosophy is unacceptable because it assumes that natural objects can act by virtue of their own essences and natures. That is, Aristotle believed in secondary causation – that physical objects can cause effects in other physical objects. As already mentioned, most theologians in Islam rejected this doctrine. It is ironic that some time before al-Ghazali, in the tenth century, opponents of natural philosophy, especially the group known as Mutazilites, used natural philosophy to attack Aristotle's doctrine of secondary causation while simultaneously defending the powerful conviction that God is the direct cause of every "natural" action. In defense of God's role as the direct cause of every effect, the Mutazilites constructed an elaborate doctrine of continuous re-creation, in which God creates the world at one moment of time, and then, by ceasing to act, allows it to vanish, only to re-create it again in the next moment of time, and again to allow it to vanish, and so on. This extraordinary interpretation was based on an atomic theory that bore little resemblance to the traditional Greek atomic theory associated with the names of Democritus and Leucippus.

Al-Ghazali found mathematics dangerous because it uses clear demonstrations, thus leading the innocent to think that all the philosophical sciences are equally lucid. A man will say to himself, al-Ghazali related, that "if religion were true, it would not have escaped the notice of these men [that is, the mathematicians] since they are so precise in this science."[6] Al-Ghazali explained further that such a man will be so impressed with what he hears about the techniques and demonstrations of the mathematicians that "he draws the conclusion that the truth is the denial and rejection of religion. How many have I seen," al-Ghazali continued, "who err from the truth because of this high opinion of the philosophers and without any other basis."[7] Although al-Ghazali allowed that the subject matter of mathematics is not directly relevant to religion, he included the mathematical sciences within the class of philosophical sciences (these are mathematics, logic, natural science, theology or metaphysics, politics, and ethics) and concluded that a student who studied these sciences would be "infected with the evil and corruption of the philosophers. Few there are who devote themselves to this study without being stripped of religion and having the bridle of godly fear removed from their heads."[8] Al-Ghazali reserved his strongest denunciation of the philosophical sciences for metaphysics, or theology, as he also called it, where "occur most of the errors of the philosophers."[9]

In his great philosophical work, *The Incoherence of the Philosophers*, al-Ghazali attacked ancient philosophy, especially the views of Aristotle, through a critique of the ideas of al-Farabi and Avicenna, two of the most important Islamic commentators on Aristotle. After criticizing their views on twenty philosophical problems, including that the world is eternal, that God knows only universals and not particulars, and that bodies will not be resurrected after death, al-Ghazali began his brief conclusion as follows:

IF SOMEONE SAYS:
    Now that you have analysed the theories of the philosophers, will you conclude by saying that one who believes in them ought to be branded with infidelity and punished with death?
*we shall answer:*
    To brand the philosophers with infidelity is inevitable, so far as three problems are concerned, namely
        (i) the problem of the eternity of the world, where they maintained that all the substances are eternal.
        (ii) their assertion that Divine knowledge does not encompass individual objects.
        (iii) their denial of the resurrection of bodies.
    All these three theories are in violent opposition to Islam. To believe in them is to accuse the prophets of falsehood, and to consider their teachings as a hypocritical misrepresentation designed to appeal to the masses. And this is blatant blasphemy to which no Muslim sect would subscribe.[10]

In al-Ghazali's view, theology and natural philosophy were dangerous to the faith. He had an abiding distrust of philosophers and praised the "unsophisticated masses of men," who "have an instinctive aversion to following the example of misguided genius." Indeed, "their simplicity is nearer to salvation than sterile genius can be."[11] As one of the greatest and most respected thinkers in the history of Islam, al-Ghazali's opinions were not taken lightly.

In his famous, and extraordinary, *Introduction to History*, Ibn Khaldun (1332–1406) includes the "physical and metaphysical sciences of philosophy," along with the religious sciences, among the "sciences that are wanted *per se*" and gave detailed descriptions of these sciences. Yet he was apparently convinced that the science of philosophy, along with the sciences of astrology and alchemy, can do great harm to religion. He expressed this sentiment in a chapter titled "A refutation of philosophy. The corruption of the students of philosophy."[12] After including al-Farabi and Avicenna among the most famous philosophers, Ibn Khaldun condemned the opinions of the philosophers as wholly wrong.[13] He informed his fellow Muslims that "the problems of physics are of no importance for us in our religious affairs or our livelihoods. Therefore, we must leave them alone."[14] Logic was potentially dangerous to the unprepared faithful. "Whoever studies it [logic]," Ibn Khaldun warned, "should do so (only) after he is saturated with the religious law and has studied the interpretation of the Qur'ân and jurisprudence. No one who has no knowledge of the Muslim religious sciences should apply himself to it. Without that knowledge, he can hardly remain safe from its pernicious aspects."[15]

I do not wish to suggest that the attitudes of al-Ghazali and Ibn Khaldun were universal among religious leaders and the educated. Islamic civilization was hardly monolithic. Unlike medieval Western Christianity, Islam had no central religious authority to impose any particular version of orthodoxy, or any particular attitude toward natural philosophy. Discussions of natural philosophy did not disappear. Nevertheless, a general uneasiness about natural philosophy seems to have been widespread. A discipline that was often viewed as at odds with the Koran was not likely to have been perceived as having significant educational value for the faithful. Its study would hardly have been encouraged, which perhaps explains why natural philosophy was never institutionalized in Islam, as it was in the Christian West.

How did such radically different attitudes toward philosophy and natural philosophy develop in the medieval civilizations of Christian Europe and Islam? Why was there such a significant split between Islamic natural philosophers and religious scholars? Why did the terms "philosophy" (*falsafa*) and "philosopher" (*faylasuf*) have less than positive connotations, especially after the impact of al-Ghazali's writings? Perhaps these difficulties for natural philosophy were to some extent consequences of the strikingly different ways in which the two religions were disseminated, and the different cultural and intellectual contexts into which they were propagated.

If Christianity was disseminated slowly, permitting centuries of adjustment to the pagan world that it would come to dominate in religion but that would eventually dominate it in philosophy, science, and natural philosophy, the religion of Islam was transmitted with remarkable speed – in about one hundred years – and over vast areas embracing many diverse peoples (see chapter 1). In contrast to Christianity, which was a relatively weak force within the Roman Empire for many centuries and therefore had to spread its ideas by missionary zeal – it had no armies to compel acceptance – Islam did not have to carry on a dialogue with the peoples it conquered and converted. Where Muslim armies prevailed, the Muslim religion was installed. Conversion to Islam was encouraged and made easy. In a short time, many, probably most, of the conquered peoples, adopted the new religion. Indeed, the conquests were to a considerable extent motivated by zeal to spread the true faith. The Muslim religion never went through a period of adjustment to pagan philosophy and learning. Thus where Christianity was born within the Roman Empire and Mediterranean civilization, and was in a subordinate position within that empire for centuries, Islam was born outside the influence of the Roman Empire and was never in a subordinate position to other religions or governments. Islam, unlike Christianity, did not have to accommodate to a wider culture or accept Greek learning, which it continued to view as alien and potentially dangerous to the Islamic faith.

*A comparison of natural philosophy in Islam and the Christian West.* Natural philosophy and theology had very different relationships in Christian Western Europe and in Islamic society. In Islam, with the exception of the *mutakallimun* and an occasional striking figure such as al-Ghazali, natural philosophers were usually distinct from theologians. Scholars were either one or the other, rarely both. Moreover, natural philosophy was always on the defensive; it was viewed as a subject to be taught privately and quietly, rather than in public, and it was taught most safely under royal patronage, as seen in the careers of some of Islam's greatest natural philosophers. Within Western Christianity in the late Middle Ages, by contrast, almost all professional theologians were also natural philosophers. The structure of medieval university education also made it likely that most theologians had early in their careers actually taught natural philosophy. The positive attitude of theologians and religious authorities toward natural philosophy within Western Christianity meant that the discipline could develop more comfortably and consistently in the West than in Islamic society. In the West natural philosophy could attract talented individuals, who believed that they were free to present their opinions publicly on a host of problems that formed the basis of the discipline.

The favorable attitude of Christianity toward natural philosophy in the West was not derived solely from the hundreds of years in which Christianity adjusted to pagan learning. Other factors also operated. Although Greco-Roman learning was suspect, it was not considered an enemy, and its potential utility was recognized early. Also, as we saw in chapter 1, a positive feeling toward natural philosophy may inadvertently have been fostered by the Christian attitude toward the state, exemplified in Jesus' call to "Render therefore unto Caesar the things which are Caesar's; and unto God the things that are God's" (Matt. 22.21). Although many Christian churchmen, such as Saint Augustine, proclaimed the superiority of the church over the state, the Christian church did acknowledge and accept the separation of church and state, which allowed for the development of a secularly oriented natural philosophy.

In medieval Islam, by contrast, truly secular government was absent, and church and state were one. The function of the state was to guarantee the well-being of the Muslim religion, so that all who lived within the state could be good, practicing Muslims. Science is essentially a secular activity. Where religion is strong, as is the case with Islam, it is very likely to dominate secular activities like science, unless the activity is recognized as independent, is protected by a secular state, or is favorably regarded by religious authorities. In medieval Islam, none of these conditions was met, whereas in late medieval Latin Christendom the third condition was clearly in effect. Because the Church looked with favor on science, secular authorities also adopted a beneficent approach toward it – they had no reason to oppose science and natural philosophy. Indeed, they found many occasions to favor the disciplines. Because the second and third conditions were fulfilled, the first was almost met as well during the late Middle Ages. Although theological constraints never vanished, they tended to be moderate and posed few obstacles to the practice of science and natural philosophy.

Certain aspects of their religion may also have drawn Christians to Greek philosophy. One example is the problem of the Eucharist, with its difficulties about the nature of substances and their attributes. Adoption of a Trinitarian position placed enormous metaphysical burdens on Christianity. Once Jesus was perceived as the Son of God, the problems of expounding the nature of the Godhead were formidable indeed. To help explain such theological difficulties, scholars deemed the concepts and terminology of Greek metaphysics essential. Logic was also considered important. Saint Augustine regularly used logic in resolving theological problems (see especially *On the Trinity* [*De trinitate*]) and in this became a model for later theologians.

Strictly unitarian religions, such as Islam and Judaism, needed no such metaphysical assistance or apparatus to expound the nature of God, although there were of course problems that seemed to require some sort of philosophical explication. For example, early on it was assumed that the Koran existed before its revelation to Muhammad. A debate arose over whether the pre-existent Koran had been created, or was uncreated and eternal. This issue, in turn, was connected to another: whether certain attributes of God mentioned in the Koran, such as "living" and "knowing," were eternal qualities independent of his essence. Nevertheless, Islam was much less driven by inherent theological needs to seek out Greek philosophy than was Christianity. Indeed, Islamic theologians discouraged analyses of the Koran and prevented the development of a speculative theology.

Latin Christianity provided a sympathetic environment for the sustenance and advance of natural philosophy and science. It posed few obstacles to their practice and development. In fact, by allowing natural philosophy to form the graduate curriculum in the medieval universities, medieval Christianity showed that it was prepared to do more than merely tolerate its existence. It actively promoted natural philosophy in an open and public way.

Even if Western Christianity had only tolerated the study and use of natural philosophy, however, that would have been a significant contribution. For we can easily imagine a scenario in which hostile religious authorities might have sought to prevent the pursuit of science and natural philosophy. Despite some difficulties in the thirteenth century, Christians avoided this course of action. Consequently, the religious attitude in Christendom favored natural philosophy more than it did in Islam, where Islam's theologians were largely hostile to it. By restricting scientific activity to those fields regarded as handmaidens to religion and theology, religious authorities dampened enthusiasm for bold investigations into nature.

Because of a fear that natural philosophy was potentially dangerous to the Muslim faith, and perhaps for other reasons as well, Islam never institutionalized natural philosophy, never made it a regular part of the educational process. Thus natural philosophy did not become part of an overall approach to nature, not even for use in determining the structure and operations of nature for the greater glory of God. It is unlikely that most Islamic thinkers thought that God's handiwork was fathomable, although, occasionally, someone like Averroes would argue that the Koran mandated that a good Muslim study nature. Averroes made this assertion in a revealing treatise, *On The Connection between Religion and Science*. He explained that the purpose of the treatise was to determine "whether the study of philosophy and logic is allowed by the [Islamic] Law, or prohibited, or commanded – either by way of recommendation or as obligatory."[16] It is remarkable that after approximately three hundred years, during which science, logic, and natural philosophy were available in Islam, Averroes felt compelled to justify their study. I know of no analogous treatise in the late Latin Middle Ages in which any natural philosopher or theologian felt compelled to determine whether the Bible permitted the study of secular subjects. It was simply assumed that it did.

The absence of an institutional base for science and natural philosophy is perhaps the most important reason why these disciplines did not become permanently rooted in Islamic society. The open hostility, or in many cases simply the lack of enthusiasm, of Islamic theologians and religious authorities provides at least one major reason why an institutional base comparable to the universities in the West failed to develop. We must not think, however, that all of Christendom was equally enthusiastic about Greek science and natural philosophy. Ironically, as will be discussed in the next section, the Byzantine Empire, the heir to the language and literature of Greek civilization, also did not make science and natural philosophy a prominent feature of education and an inherent part of its culture.

By contrast, the universities that were founded in the West in the Middle Ages served to preserve and enhance natural philosophy. As we have seen, the university as we know it today was invented in the late Middle Ages. Universities were powerful and highly regarded institutions, corporate entities with numerous privileges that increased century by century. Despite plagues, wars, and revolutions, they carried on, giving to natural philosophy and science continuity and permanence. They could do so because the Church, and the theologians who were the guardians of Church doctrine, had acquiesced to Aristotelian natural philosophy's having a major role in education. For the first time in history, science and natural philosophy had a permanent institutional base. No longer would the preservation of natural philosophy be left to the whims of fortune and to the efforts of isolated teachers and students.

Before 1500, the exact sciences in Islam had reached lofty heights, greater than they achieved in medieval Western Europe, but they did so without a vibrant natural philosophy. In contrast, in Western Europe natural philosophy was highly developed, whereas the exact sciences were merely absorbed (from the body of Greco-Arabic scientific literature) and maintained at a modest level. After 1500, Islamic science effectively ceased to advance, but Western science entered upon a revolution that would culminate in the seventeenth century. What can we learn from this state of affairs?

Let me propose the following: that the exact sciences are unlikely to flourish in isolation from a well-developed natural philosophy, whereas natural philosophy is apparently sustainable at a high level even in the absence of significant achievement in the exact sciences. One or more of the exact sciences, especially mathematics, was practiced in a number of societies that never had a fully developed, broadly disseminated natural philosophy. In none of these societies had scientists attained as high a level of competence and achievement as they had in Islam. Was then the subsequent decline of science in Islam perhaps connected with the relatively diminished role of natural philosophy in that society and to the fact that it was never institutionalized in higher education? This is a distinct possibility, if natural philosophy played as important a role as I attribute to it throughout this study. Thus in Islamic society, where religion was so fundamental, the absence of support for natural philosophy from theologians, and, more often, their open hostility toward the discipline, might have proved fatal to it and, eventually, to the exact sciences as well.

*The other Christianity: Science and natural philosophy in the Byzantine Empire.*
Up to this point, I have treated Christianity as a single, unified entity. In truth, we have only discussed the Latin Christianity of Western Europe. Now, we must briefly examine the eastern part of the Roman Empire, which developed a brand of Christianity that differed considerably from its western counterpart. At the outset of the Roman Empire, the eastern, or Byzantine, part and the western, or Latin, part formed one unified state. Within this unified empire, Christianity was essentially one. As time passed, however, the Roman Empire became two distinct, and even rival, subdivisions (see chapter 1). By 800, the old Roman Empire was de facto divided into western and eastern parts, each ruled independent of the other. The split was manifested linguistically. In the West, Latin was the common and official language; in the East, it was Greek.

As might be expected, the division of the empire also affected religion. Christianity split into two rival factions, the Catholic Church in the West, and the Greek Orthodox Church in the East (and, to a small extent, in certain areas of the West). They differed in their use of liturgical languages, with Latin being used in the West, and Greek in the East. The Eastern clergy were allowed to marry, the Western clergy were not. In the Holy Eucharist, or mass, clergy in the West used unleavened bread, whereas the Eastern clergy used leavened bread. In the Eastern church, laymen could be appointed as patriarchs. During the course of the Byzantine Empire, this extraordinary practice – it was unheard of in the Western church – was utilized thirteen times in the selection of 122 patriarchs of Constantinople. The most important difference, however, occurred in the early sixth century, when the Catholic Church changed the

Nicene Creed of 325 A.D. Where the latter had formerly declared that the Holy Spirit proceeded "from the Father" alone, the Catholic Church added the words "and the Son" (*filioque*). The Holy Spirit was now said to proceed from both the Father and the Son, a claim that the Greek church found objectionable, because it seemed to say that the Holy Spirit was derived from two distinct Gods. The formation of two distinct Christian churches was a de facto reality long before 1054, when papal legates on a mission to Constantinople excommunicated the patriarch and his associates, who, in turn, condemned the papal legates.

One of the major features of Western Christianity was its differentiation from the secular state. Each had its role to play and although, at different times, church and state sought to dominate each other, they usually recognized their mutually established spheres of jurisdiction. By contrast, Byzantium was essentially a theocratic state. The distinction between church and state was largely non-existent. The emperor was regarded as the viceroy of God and a sacred leader. No fierce debate raged in the Byzantine Empire about the relative merits and powers of secular versus spiritual authorities, as occurred in the West. The Byzantine emperor not only made all secular decisions autocratically, but he also exercised near total control over the administration of the Greek church. He could, for example, appoint and depose the patriarchs of the church. On occasion, the emperors even tried to change church dogma and the sacraments, although never successfully.

Compared to their counterparts in the Latin West and Islam, the inhabitants of the Byzantine Empire were privileged. Because Greek was their native language, they could read Greek works on science and natural philosophy directly, without need of translation. Moreover, these Greek works of science and natural philosophy were readily available in the libraries and depositories of Constantinople and the surrounding region. Indeed, most of our manuscript sources for Greek science have come from the Byzantine Empire. In Byzantium, an educated elite had always been instructed in the Greek classics, reading, among other works, the *Iliad*, and authors such as Hesiod, Pindar, Aristophanes, and the Greek tragedians. Byzantium valued both public and private education, which were largely in the hands of lay authorities, rather than the Church. As evidence of this high regard are the words of Theodore II, emperor of the truncated Byzantine state that was governed from Nicaea following the sack of Constantinople in 1204, who declared, "Whatever the needs of war and defence, it is essential to find time to cultivate the garden of learning."[17]

Despite these intellectual advantages, scholars in Byzantium failed to capitalize on their good fortune. "The garden of learning" seems to have produced few flowers for the history of science and natural philosophy. Any judgment on the achievements of Byzantium in these areas of learning must, however, be qualified by the realization that the greatest part of the literature on these subjects lies as yet unpublished and therefore, for the most part, unread.

The most significant scientific accomplishments occurred relatively early in the history of the Byzantine Empire, primarily during the fourth to sixth centuries, when noteworthy contributions were made in mathematics and engineering and when Greek commentators, such as Themistius, Simplicius, and John Philoponus, wrote Aristotelian commentaries that eventually exerted influence in Islam and the Latin West. But after the emperor Justinian ordered the closing of Plato's Academy in Athens in 529 on religious grounds, Byzantium would honor the ancient Greek tradition only sporadically. Until the fall of the Byzantine Empire in 1453, at least two identifiable renaissances in learning occurred, once in the eleventh century and again during the empire's last two centuries.

Why during the last 800 years of the Byzantine Empire did scholars add little of significance to their enormous legacy of ancient Greek science and natural philosophy? Why did the civilizations of Islam and Western Europe outperform Byzantium by large margins, despite a total reliance on translations, and with fewer ancient works available?

One might, at first glance, think that war played a significant role in minimizing intellectual activity in the Byzantine Empire. Although wars afflict most societies and civilizations, Byzantium was different. It was constantly at war defending an ever-shrinking empire that lasted more than a thousand years only because the empire was prepared to make huge sacrifices to preserve itself. Throughout its history, Byzantium was surrounded by enemies, who were attracted by its enormous wealth. To the east, the Byzantines fought first Persians and later Arabs; to the north, they battled Slavs and Turkic tribes (especially the Seljuk Turks, who dealt them a severe blow in 1071 at Manzikert in Armenia). The onslaught against them was relentless. Yet even after its loss of Asia Minor to the Seljuk Turks, the empire survived for nearly four more centuries. Ironically, the cruelest blow against the Byzantines, and perhaps ultimately the fatal one, was delivered by their fellow Christians from the West, who, during the Fourth Crusade, in 1204, decided to capture and sack Constantinople. A noted historian of the Byzantine Empire has remarked that "in its long history Byzantium managed to hold off enemy after enemy, a power of resistance due not only to its superior military tactics, the great defensive walls of Constantinople, and the stability of its currency, but also to the steadfast belief of its people in the purity of their Orthodox religious faith and the divine origin of their empire."[18] We might readily believe that in a civilization constantly at war, or preparing for war, intellectual activities were easily neglected. Although wars might have disrupted intellectual activities in Byzantium on occasion, such an interpretation, however, would be inaccurate because the Byzantine Empire experienced its greatest intellectual renaissance during the two final, desperate, war-filled centuries of its existence.

Was the Orthodox Church in Byzantium an obstructionist force in science and natural philosophy? It did occasionally interfere (for example, it opposed the emphasis on pagan learning in the eleventh-century renaissance), but was probably not the major factor in Byzantium's minuscule achievements in science and natural philosophy. Although the Orthodox Church discouraged the study of pagan literature as much as it could, it was well aware that it could not wholly prevent it. And when it was convenient, the Church itself drew upon the ancient Greek traditions. But it never had an easy relationship with pagan thought. We saw earlier that the handmaiden concept in the West was gradually ignored and eventually abandoned. Its fate in the East was quite different. Throughout the history of the Byzantine Empire, Church authorities insisted on the handmaiden approach to philosophy and to secular learning in general, showing hostility toward any attempt to study such subjects either for their own sakes or for the sheer love of knowledge. The Church's theologians were only occasionally natural philosophers. Perhaps the empire's almost permanent state of war narrowed the focus of the Eastern Church and led it to distrust anything that might undermine the faith and its own authority.

Nevertheless, as we have seen, it is ironical that in the last two centuries of Byzantium's existence, Byzantine learning and scholarship underwent a renaissance, even as the empire was descending into destruction. "If there is any meaning in the concept of decadence," wrote Sir Steven Runciman, a distinguished Byzantinist, "there are few polities in history that better deserve to be called decadent than the East Christian Empire, the once great Roman Empire, during the last two centuries of its existence."[19] And yet, even as the state was crumbling, it experienced a great surge of interest in philosophy and science and in learning in general. Most of the significant names of this renaissance, who wrote on philosophy and science – scholars such as Gregory Choniades, George Chrysococces, George Acropolites, the emperor Theodore II, George Pachymer, and Maximus Planudes – are unknown to historians of science and philosophy. Two scholars who not only wrote on these subjects but were also wealthy patrons of other scholars were Nicephorus Chumnus and Theodore Metochites. Perhaps the best-known name among these scholars is that of George Gemistus Plethon, who lectured on Plato in Florence, where he had been sent as a delegate to the Council of Ferrara-Florence (1439). Indeed, Plethon found more honor in Italy than in Greece, as did his most distinguished student, Cardinal Bessarion, who abandoned the Orthodox faith to become a cardinal of the Roman Catholic Church.

This last renaissance within the Byzantine Empire was, of course, doomed. As a modern scholar has put it: "They [Byzantine scholars] wrote for the most part in a sophisticated language for a sophisticated public which was soon to be wiped out."[20] With the fall of Constantinople in 1453, it came to an abrupt, inglorious end. What was the overall impact of this unusual renaissance, as well as of the Byzantine scholarship that had preceded it as far back as the seventh century? No effect of Byzantine scholarship has been detected beyond the boundaries of the empire itself. And, although some significant thoughts may be embedded in the vast literature, most of it unpublished, that we still have, future research is unlikely to discover much that will alter modern opinion about Byzantine thought and scholarship, which have been characterized as uninspired and unoriginal. Even if a few important, innovative works are eventually discovered, we can be reasonably certain that they had no impact on the course of scholarship and learning beyond Byzantium. They had simply been unknown. At best, such discoveries would inform the world that some Byzantine scholars had been capable of significant achievements, and might have added to the sum total of learning, if only their works had been disseminated.

The character of Byzantine scholarship was revealed in the preface to Theodore Metochites's *Historical and Philosophical Miscellanies* (*Miscellanea Philosophica et Historica*). Metochites spoke for Byzantine philosophers when he declared, "The great men of the past have said everything so perfectly that they have left nothing for us to say."[21] We should contrast this attitude with that of scholars in Islam and the Latin West, who also respected the ancients, but who were often prepared to go beyond them and add to the sum total of knowledge. Byzantine scholarship stands in sharp contrast to the questions tradition in the West, where authors were forced to confront one issue after another and to devise reasoned responses that might, or might not, agree with the author of the text.

In the Byzantine Empire, scholarship – and therefore scholarship in science and natural philosophy – was done primarily by a tiny minority of laymen, who shared little other than a common educational background. They certainly did not reflect deeply on a wide range of common problems, in contrast to scholars at the universities in the West. It seems that Byzantine scholarship was formalistic and pedantic, rarely innovative. As Donald Nicol, the Byzantinist, has described it:

> The prime purpose of scholarship as practised in late Byzantium seems often to have been the gratification of one's own ego and sometimes the discomfiture of one's intellectual opponents. . . . Beyond the discipline of the master-pupil relationship there was little observable cooperation among Byzantine scholars. Each worked alone as an individualist. He would often express his debt to the giants of antiquity – to Plato, Aristotle or Ptolemy. But he would seldom acknowledge the work of any of his

immediate predecessors, still less that of one of his contemporaries. . . . Certainly the scholars congratulated each other on their expertise in their elaborate correspondence. Flattery was part of rhetoric. Sometimes they borrowed books from each other. But only rarely did they collaborate in their researches.[22]

Thus although Byzantine scholars engaged in self-congratulatory scholarship and flattery, they seem to have had little desire to resolve problems in natural philosophy. Perhaps the handmaiden approach was also somewhat inhibiting, as in the example of the ill-fortune that befell John Italos in 1082. Italos, a professor of philosophy in Constantinople, was declared a heretic because he came to love Hellenic learning too much. His mistake was widely known because "his errors were read out in every Orthodox church on the first Sunday of Lent as a perpetual reminder and warning."[23] When this is added to the uncritical nature of Byzantine scholarship, wherein most scholars who commented on ancient texts were convinced that the ancients were always right and therefore believed that they could add little or nothing, the unremarkable character of their literature in science and natural philosophy is understandable.

The apparent absence of more penetrating and innovative scholarship is also attributable to the fact that neither church nor state – and they were often one and the same – ever institutionalized the study of natural philosophy and science. In this, Byzantium differed little from Islam. The theologians in both civilizations were either hostile or indifferent to science and natural philosophy. Islam, however, treated the initially alien disciplines of Greek science and natural philosophy more enthusiastically and successfully than Byzantium treated its own native Greek heritage. The civilization of Islam transmitted much new scientific knowledge to the West along with the Greek classics. Byzantium passed virtually no new science and philosophy of its own making to the West.

Although it is important to know why Byzantine scholars were not, as it seems, intellectually productive in natural philosophy and science, it is more vital and appropriate to recognize that their real intellectual significance lies in the preservation and transmission of the Greek scientific tradition. For this incalculable contribution, the Byzantines have rightly been called "the world's librarians" in the European Middle Ages.[24] In this sense, the Byzantine Empire played a monumental role in the history of science and learning.

## THE SUBSTANTIVE PRE-CONDITIONS THAT MADE THE SCIENTIFIC REVOLUTION POSSIBLE

Without the development of the three contextual pre-conditions already described, it is difficult to imagine how a scientific revolution could have occurred in the seventeenth century. Although these pre-conditions, which were permanent features of medieval society, were vital for the emergence of early modern science, and therefore qualify as foundational elements, they were not in themselves sufficient. The reasons why science took root in Western society must ultimately be sought in the nature of the science and natural philosophy that were developed. It is therefore also essential to examine the basic features of medieval science and natural philosophy. What about them justifies the claim that the Middle Ages contributed to generating the new science that emerged in the seventeenth century?

## The exact sciences

If we leave medicine aside, science in the Middle Ages is appropriately divisible into the exact sciences (primarily mathematics, astronomy, statics, and optics) and natural philosophy. I shall concentrate on natural philosophy. Although the Latin Middle Ages preserved the major texts of the exact sciences, and even added to their total, I am not aware of methodological or technical changes derived from these works that proved significant for the Scientific Revolution. Preserving the texts and studying them, and even writing new treatises on these subjects, were themselves significant achievements. Not only did these activities keep the exact sciences alive, but they also revealed a group of individuals who, during the medieval centuries, were competent to do work in these sciences. At the very least, expertise in the exact sciences was maintained, so that the Copernicuses, Galileos, and Keplers of the new science had something that they might study and, perhaps, alter for the better. Indeed, what would they have done without them? Nevertheless, during the Middle Ages, there was considerably more innovation in natural philosophy than in the exact sciences. Medieval natural philosophy shaped what was to come more than did the exact sciences.

## Natural philosophy: The mother of all sciences

The role of natural philosophy during the Middle Ages differed radically from that of the exact sciences. With natural philosophy, we are not concerned with the mere preservation of Greco-Arabic knowledge, but rather with the transformation of an inheritance into something ultimately beneficial for the development of early modern science. In chapter 7, I described the essential features of natural philosophy. Now it is time to determine its relevance for the dramatic changes in science that occurred in the sixteenth and seventeenth centuries.

I have not yet discussed one important characteristic of natural philosophy because, until now, it was irrelevant. In the larger context of the relationship of natural philosophy to the Scientific Revolution, however, it assumes great significance. I refer to the concept of natural philosophy as the mother of all sciences. By the time Aristotelian natural philosophy reached Western Europe, astronomy, mathematics, geometrical optics, and medicine had been independent sciences in the Greco-Arabic tradition for a long time. As "middle sciences," astronomy and optics were intermediate between mathematics and natural philosophy and were occasionally thought of as part of the latter. Medicine, always intimately associated with natural philosophy, was also an independent discipline and had been so since at least the fifth century B.C. As acknowledgment of its independence, medicine had its own faculty in the medieval university.

If we exclude astronomy, mathematics, optics, and medicine, almost all other sciences – physics, chemistry, biology, geology, meteorology, and psychology, as well as all their subdivisions and branches – emerged as independent disciplines from within the matrix of natural philosophy during the seventeenth to nineteenth centuries. They emerged only slowly, and most were not recognized as distinct subjects until the nineteenth century. By the mid-nineteenth century natural philosophy was still widely taught in colleges in the United States. A definition of it given for a course of lectures at Dickinson College (Pennsylvania) in 1845–1846 declared that

> *Natural philosophy* may be considered as the science which examines the general and permanent properties of bodies; the laws which govern them, and the reciprocal action which these bodies are capable of exerting upon each other, at greater or less distances, without changing their matter.[25]

This definition considerably narrows the scope of the medieval conception of natural philosophy, and it is also remarkably different because "at mid-century, the divisions of natural philosophy were mechanics, hydrostatics, hydrodynamics, hydraulics, pneumatics, acoustics, optics, astronomy, electricity, galvanism, magnetism, and chromatics."[26] It is obvious that by 1850, approximately six centuries after natural philosophy had become the curriculum in the medieval universities, the "mother of all sciences" had given birth to her numerous progeny, many of whom were apparently reluctant to leave her congenial abode to dwell in as yet new and unfamiliar surroundings. It is, however, strange to see optics and astronomy back with "mother" after having lived more or less apart during the Middle Ages.

The natural philosophy that spawned these sciences over many centuries was developed in the universities of Western Europe during the late Middle Ages. It was a unique development in the history of the human race. Natural philosophers in the arts faculties of the universities converted Aristotle's natural philosophy into a large number of questions that were put to nature on a range of subjects that would eventually crystallize into the sciences just mentioned. To each of these questions two or more conclusions were usually provided. Revolutionary changes occurred when the conclusions that were acceptable to natural philosophers in the Middle Ages were found inadequate by scholars in the sixteenth and seventeenth centuries. By the end of the seventeenth century, new conceptions of physics, and of the cosmos as a whole, drastically altered natural philosophy. Aristotle's cosmology and physics were largely abandoned. But his ideas about many other aspects of nature, including, for example, material change, zoology, and psychology, were still found useful. In biology, Aristotle's influence continued into the nineteenth century.

Although medieval natural philosophy was transformed into something that subsequently spawned a host of new independent, scientific disciplines, Aristotelian natural philosophy had already been significantly altered in the fourteenth century. That transformation undoubtedly played a role in the revolution to come. But this was not because of any particular achievements in science, although these are by no means negligible. Medieval natural philosophers were interested in the ways that we can know and approach nature, that is, in what might be called scientific method. They sought to explain how we come to understand nature, even though they rarely pursued the consequences of their own methodological insights. The methodological achievements of medieval natural philosophers have already been described (see chapter 7). They formed a significant part of the medieval legacy to the early modern world and deserve brief mention in the present context.

A few of these methodological changes were relevant to mathematics. In developing his new mathematical description for motions produced by the application of motive forces to resisting bodies, Thomas Bradwardine departed from Aristotle when he realized that natural processes had to be represented by mathematical functions that hold for all values and are therefore continuous. Scholastic authors frequently introduced infinites and infinitesimals into nature and were aware that one could treat the infinitely large and the infinitely small in the same manner as one treated finite quantities.

The mathematical treatment of qualities was characteristic of medieval natural philosophy. Although the problems were usually imaginary and hypothetical, the application of mathematics to solve such problems was commonplace. By the sixteenth and seventeenth centuries, mathematical ways of thinking, if not mathematics itself, had been incorporated into natural philosophy. The stage was set for the consistent application of natural philosophy to real physical problems, rather than to the ways in which qualities were imagined to vary.

Most of the methodological contributions to science were, however,

philosophical. Scholastic natural philosophers formulated sound interpretations of concepts such as causality, necessity, and contingency.
Some scholars – and John Buridan, the eminent arts master at the University of Paris in the mid-fourteenth century, was one of them – concluded that final causes were superfluous and unnecessary. For them,
efficient causes were sufficient to determine the agent of a change. Buridan also addressed another major methodological development when
he insisted that scientific truth is not absolute like mathematical truth is,
but has degrees of certitude. The kind of certainty Buridan had in mind
consisted of indemonstrable principles that formed the basis of natural
science, as, for example, that all fire is warm and that the heavens move.
For Buridan, these principles are not absolute but are derivable from
inductive generalization, or, as he put it, "they are accepted because they
have been observed to be true in many instances, and to be false in
none."[27]

As we saw in chapter 7, Buridan regarded these inductively generalized principles as conditional, because their truth is predicated on the
"common course of nature," a profound assumption that effectively
eliminated the impact on science of unpredictable, divine intervention.
In short, it eliminated the need to worry about miracles in the pursuit
of natural philosophy. It also neutralized chance occurrences that
might occasionally impede or prevent the natural effects of natural
causes. Just because individuals are occasionally born with eleven fingers does not negate the fact that in the common course of nature we
can confidently expect ten fingers. On this basis, Buridan proclaimed
the possibility of truth with certitude. Using reason, experience, and
inductive generalization, Buridan sought to "save the phenomena" in
accordance with the principle of Ockham's razor – that is, by the simplest explanation that fit the evidence. Buridan only made explicit
what had been implied by his scholastic colleagues. The widespread
use of the principle of simplicity was a typical feature of medieval natural philosophy, and was characteristic of science in the seventeenth
century, as when Johannes Kepler declared that "it is the most widely
accepted axiom in the natural sciences that Nature makes use of the
fewest possible means."[28]

Thus medieval natural philosophers sought to investigate the "common course of nature," not its uncommon, or miraculous, path. They
characterized this approach admirably by the phrase "speaking naturally" (*loquendo naturaliter*) – that is, speaking in terms of natural science,
and not in terms of faith or theology. That such an expression should
have emerged and come into common usage in medieval natural philosophy is a tribute to the scholars who took as their primary mission
the explanation of the structure and operation of the world in purely
rational and secular terms.

The widespread assumption of "natural impossibilities," or counterfactuals, or, as they are sometimes called, "thought experiments," as described in chapters 5 and 7, was a significant aspect of medieval methodology. A hypothetical occurrence would have been considered "naturally impossible" if it were thought inconceivable within the accepted framework of Aristotelian physics and cosmology. Natural impossibilities were derived largely from the concept of God's absolute power as embodied in the Condemnation of 1277. Counterfactuals allowed the imagination to soar. In the Middle Ages, such thinking resulted in conclusions that challenged aspects of Aristotle's physics. Whereas Aristotle had shown that other worlds were impossible, medieval scholastics showed not only that other worlds were possible but that they would be compatible with our world. The novel replies that emerged from the physics and cosmology of counterfactuals did not cause the overthrow of the Aristotelian world view, but they did challenge some of its fundamental principles. They created an awareness that things could be quite different than was dreamt of in Aristotle's philosophy. But they accomplished more than that. Besides influencing scholastic authors in the sixteenth and seventeenth centuries, this characteristically medieval approach also influenced important nonscholastics, who were aware of the topics debated by scholastics.

One of the most fruitful ideas that passed from the Middle Ages to the seventeenth century was that God could annihilate matter and leave behind a vacuum. For example, John Locke based his argument for the existence of a three-dimensional void space on the assumption that God could annihilate any part of matter. If God did destroy a body, a vacuum would remain, "for it is evident that the space that was filled by the parts of the annihilated body will still remain, and be a space without a body."[29] In a similar, though somewhat more complicated, manner, Pierre Gassendi (1592–1655), a French philosopher, also concluded that an infinite, three-dimensional void space existed. He demonstrated the validity of this claim in stages, by imagining the supernatural annihilation of all matter first within the sublunar region, then in the celestial region beyond the moon, and, finally, in a world that he imagined was becoming successively larger and larger. For "if there were a larger world, and a larger one yet, on to infinity, God successively reducing each of them equally to nothingness, we understand that the spatial dimensions would always be greater and greater, on to infinity," so that we can "likewise conceive that the space with its dimensions would be extended in all directions into infinity."[30] The same principle could be applied within the world. Gassendi explained that

it is frequently necessary to proceed in this fashion in philosophy, as when they tell us to imagine matter without any form in order to permit us to understand its nature. . . . Therefore there is nothing that prevents us from supposing that the entire region contained under the moon or between the heavens is a vacuum, and once this supposition is made, I do not believe that there is anyone who will not easily see things my way.[31]

Thomas Hobbes (1588–1679), the great English philosopher, also made the annihilation of matter a principle of analysis. Despite his omission of God as the annihilator, Hobbes paid unwitting tribute to his scholastic predecessors when he declared, "In the teaching of natural philosophy, I cannot begin better (as I have already shewn) than from *privation*; that is, from feigning the world to be annihilated," a process that, among other things, enabled Hobbes to formulate his concepts of space and time.[32]

A significant natural impossibility that derived from the Condemnation of 1277 involved article 49, which made it mandatory after 1277 to concede that God could move the world rectilinearly, despite the vacuum that might be left behind. More than an echo of this imaginary manifestation of God's absolute power reverberated through the seventeenth century. When Gassendi declared that "it is not the case that if God were to move the World from its present location, that space would follow accordingly and move along with it," he was using the supernatural motion of the world as a convenient support for his belief in the absolute immobility of infinite space.[33] As spokesman for Isaac Newton, Samuel Clarke (1675–1729), in his dispute with Leibniz, also defended the existence of absolute space when he argued that "if space was nothing but the order of things coexisting [as Leibniz maintained]; it would follow that if God should remove the whole material world entire, with any swiftness whatsoever; yet it would still always continue in the same place."[34] Finally, the power of counterfactuals is nowhere more impressively illustrated than in the principle of inertia, which Newton proclaimed as the first law of motion in *The Mathematical Principles of Natural Philosophy* (1687): "Every body continues [or perseveres] in its state of rest, or of uniform motion in a right line, unless it is compelled to change that state by forces impressed upon it."[35] In medieval intellectual culture, where observation and experiment played negligible roles, counterfactuals were a powerful tool, because they emphasized metaphysics, logic, and theology, the very subjects in which medieval natural philosophers excelled.

The scientific methodologies described here and in chapter 7 would not, if followed, have produced new knowledge. No methodologies are likely to do that. But they did produce new conceptualizations and assumptions about nature and our world. Galileo and his fellow scientific revolutionaries inherited these attitudes to which most of them would have subscribed.

*Medieval natural philosophy and the language of science*

Another legacy from the Middle Ages to early modern science was an extensive and sophisticated body of terms that formed the basis of scientific discourse. These included "potential," "actual," "substance," "property," "accident," "cause," "analogy," "matter," "form," "essence," "genus," "species," "relation," "quantity," "quality," "place," "vacuum," and "infinite." These Aristotelian terms were a significant component of scholastic natural philosophy. But the language of medieval natural philosophy did not consist solely of translated Aristotelian terms. New concepts, terms, and definitions were added, most notably in the domains of change and motion. Medieval natural philosophers distinguished between dynamics, or the causes of motion, and kinematics, or the spatial and temporal effects of motion. They also differentiated between the measure of the intensity of a quality (*latitudo qualitatis*) and the total quantity of the quality distributed over an entire subject (*longitudo qualitatis*). For example, medieval natural philosophers distinguished between the intensity of heat (temperature) and the quantity of heat and between total weight (an extensive factor) and specific weight (an intensive factor). They also devised an impressive series of new kinematic definitions that were important in the history of physics, including definitions of uniform motion (*motus uniformis*), uniform acceleration (*motus uniformiter difformis*), and instantaneous velocity (*velocitas instantanea*). In dynamics, they employed the concept of impressed force, or impetus, as it was often called, which continued to play a role in physics through most of the seventeenth century. By the latter part of the century, these terms, concepts, and definitions were embedded in the language and thought processes of European natural philosophers.

*Medieval natural philosophy and the problems of science*

Medieval natural philosophy played another critical role in the transition to early modern science. It furnished at least some of the basic problems that exercised the minds of non-scholastic natural philosophers in the sixteenth and seventeenth centuries. To illustrate how medieval problems influenced developments in the new science, we need only mention our earlier discussion of Galileo's treatment of motion in chapter 6 and recall that he was concerned with traditional medieval problems of motion in plenum and void. The resolution of those problems lay at the heart of the new physics that Galileo constructed. Problems about the vacuum were also an inheritance from the Middle Ages, as were numerous problems about the terrestrial and celestial regions. The revolution in physics and cosmology was not the result of new questions put to nature in place of medieval questions. It was, at least initially, more a matter of finding new answers to old questions, answers that sometimes involved experiments, which were exceptional occurrences in the Middle Ages.

The medieval principle that "nature abhors a vacuum" provides a good illustration. Numerous experiences were cited in which this principle was invoked to explain why a vacuum could not occur. Among these were a burning candle in an enclosed vessel, the rise of water in a siphon, the rise of blood in a cupping glass, the clepsydra (water clock), and the separation of two plane surfaces. Despite differences, all of these phenomena were explained traditionally by nature's abhorrence of a vacuum. Matter would behave in strange, and even in naturally impossible, ways to prevent a vacuum. In the sixteenth century, however, many natural philosophers explained the very same phenomena by the assumption of artificially created vacua. What had previously been denied was now affirmed. Thus did old problems receive new answers.

Medieval natural philosophers posed hundreds of specific questions about nature, and their proposed answers to these questions included a massive amount of scientific information. Most of these questions came with multiple answers and offered no way of choosing the best one. In the sixteenth and seventeenth centuries, new solutions were proposed by scholars who deemed Aristotelian answers unacceptable. Their changes were made mostly in the answers, not in the questions. Although the solutions differed, many of the fundamental problems were common to medieval scholars and to scholars in the period of the Scientific Revolution that followed. Beginning around 1200, medieval natural philosophers, located largely at the universities, exhibited great concern for the nature and structure of the physical world. The contributors to the Scientific Revolution continued this tradition, which by then had become integral to intellectual life in Western society.

### Freedom of inquiry and the autonomy of reason

The Middle Ages transmitted not only a massive amount of traditional, natural philosophy that had evolved over the centuries, much of it in questions form, but also a remarkable legacy of relatively free rational inquiry. Medieval philosophical tradition was fashioned in the faculties of arts of the universities. Almost from the outset, masters of arts struggled to win as much academic freedom as they possibly could. They sought to preserve and expand the study of philosophy. Arts masters regarded themselves as the guardians of the discipline, and they fought for the right to apply reason to problems about the physical world. Because of their independent status as a faculty, with numerous rights and privileges, they achieved a surprisingly large degree of freedom during the Middle Ages.

Although theology was always a potential obstacle to the study of natural philosophy, theologians themselves offered little opposition to the discipline, largely because they too were heavily involved with it. Albertus Magnus, the eminent medieval theologian, regarded natural philosophy as independent of theology. By the middle of the thirteenth century, he considered himself to be sufficiently expert in theology to offer responsible opinions in that field, but he acknowledged that in physics he trusted "the opinion of Peripatetics" more than he did his own.[36] By the end of the thirteenth century, univerity arts faculties had attained virtual independence from theological faculties. The Condemnation of 1277 was the last significant effort by theologians to inhibit arts masters in their pursuit of natural philosophy. Some of the articles condemned by the theologians inadvertently reveal what many natural philosophers thought about their role. Among the most interesting are

40. That there is no more excellent state than to devote oneself to philosophy.
145. That there is no question disputable by reason which a philosopher should not dispute and resolve. . . .
150. That a man ought not to be satisfied to have certitude on any question on the basis of authority.

It is not difficult to see why theologians were disturbed if such opinions were widely held by natural philosophers. Philosophy is exalted over theology; and authority, including church and biblical authority, is rejected as the basis of any conclusive argument. Only reasoned argument is acceptable. We can see that if natural philosophers had been free to establish their own disciplinary criteria on the basis of such sentiments, theology would have had little role, if any. Even if theology had played a role, reason would have been applied to revelation, with disastrous consequences, as eventually happened in the seventeenth and eighteenth centuries. The ideal of natural philosophy was to use only reasoned arguments. Medieval natural philosophy was quintessentially rational.

If the lofty goals of natural philosophers were not fully realized during the Middle Ages, the path toward their realization was clearly laid in that period. By the end of the thirteenth century, philosophy, including its major subdivision, natural philosophy, had emerged as an independent discipline in the universities. Although arts masters were always restrained by religious dogma, the subject areas in which doctrinal issues arose were limited. During the thirteenth century, arts masters had learned how to cope with the problematic aspects of Aristotle's thought – they either treated problems hypothetically, or they announced that they were merely repeating Aristotle's opinions, even as they elaborated on his arguments. During the Middle Ages, natural philosophy remained what Aristotle had made it: an essentially secular and rational discipline.

It remained so only because the arts faculties struggled to preserve it. In the process, they transformed natural philosophy into an independent discipline that had as its objective the rational investigation of all problems relevant to the physical world. In the 1330s, William of Ockham expressed the sentiments of most arts masters and many theologians about natural philosophy as an independent and rational discipline when he declared that: "Assertions . . . concerning natural philosophy, which do not pertain to theology, should not be solemnly condemned or forbidden to anyone, since in such matters everyone should be free to say freely whatever he pleases."[37]

The spirit of free inquiry in natural philosophy was established in the arts faculties during the Middle Ages. Scholars supplied the best answers they could within the limits of their discipline. In chapter 7, I described varied approaches to Aristotle's authority. It should be emphasized, however, that natural philosophers considered it their duty to use reason, not faith, in their arguments. Theologians writing on natural philosophy usually adopted this approach. Although aware of the difficulties in Aristotle's thought with regard to faith, they recognized that appeals to faith did not constitute an argument. Nicole Oresme assumed this attitude when he invoked reason to repudiate Aristotle's arguments for the eternity of the world. As he expressed it: "I want to demonstrate the opposite according to natural philosophy and mathematics. In this way it will become clear that Aristotle's arguments are not conclusive."[38] Leaving aside theology and faith, Aristotle's solutions were sometimes criticized for inadequacy. For example, in the prologue to his questions on the *Physics*, John Buridan explained that in almost every question raised, Aristotle's solution was unsatisfactory. And as mentioned in chapter 7, Nicole Oresme suggested that Aristotle may have misunderstood the rules of motion that he himself had proposed in the seventh book of his *Physics*.

Scholastics were also aware of their own limitations. Buridan revealed his considerable uncertainty in the case of one difficult question when he proclaimed that "because of the multitude of probable arguments on each side and because of a deficiency of clearly demonstrative arguments,"[39] he should opt for neither side. Elsewhere, he proposed his own explanations, which must suffice "unless someone should find a better one."[40] Such doubts do not appear often, but just enough to remind us that medieval natural philosophers were at least aware of the difficulties of providing an answer to every question. Despite some hesitation, however, they considered it their obligation to inquire freely into all manner of questions, and to provide answers.

Students of natural philosophy in the sixteenth and seventeenth centuries were the beneficiaries of the spirit of free inquiry nurtured by medieval natural philosophers. Yet most were unaware of their legacy and would probably have denied its existence, preferring to ridicule Aristotelian scholastics and scholasticism. Their criticism was not without justification. The time had come to alter the course of natural philosophy. Some scholastic natural philosophers tried to accommodate the new heliocentric astronomy that had emerged from the efforts of Copernicus, Tycho Brahe, and Galileo. But accommodation was no longer enough, and medieval natural philosophy was totally transformed in the seventeenth century. The medieval scholastic legacy remained, however, in the form of the spirit of free inquiry, the emphasis on reason, the variety of approaches to nature, and the core of problems to be studied. The new science also inherited from the Middle Ages the profound sense that seeking to discover the way the world operated was a laudable undertaking.

Peter Dear, a historian of seventeenth-century science, has identified six essential and innovative features of the new science that came to dominate Western civilization in the seventeenth century.[41]

1. deliberate and recordable experimentation;
2. the acceptance of mathematics as a privileged tool for disclosing nature;
3. the reassignment of the causes of certain perceived attributes of things from the things themselves to the perceptual apprehensions of the observer (the "primary and secondary quality" distinction);
4. the associated plausibility of seeing the world as a kind of machine;
5. the idea of natural philosophy as a research enterprise rather than as a body of knowledge;
6. and the reconstruction of the social basis of knowledge around a positive evaluation of cooperative research.

Although this was indeed a "new world view," these innovations would have been impossible without the foundation laid for them in the Middle Ages. Before scholars could have an "idea of natural philosophy as a research enterprise," they first had to have a natural philosophy that deserved to be characterized as a "body of knowledge." Such a body of knowledge was fashioned in the late Middle Ages. Some of the other "modern" departures on Dear's list were not unknown in the Middle Ages. Occasional experiments had been made, and mathematics had been routinely applied to hypothetical, though rarely real, problems in natural philosophy. In the seventeenth century, the new scientists applied mathematics to real physical problems and added experiments to the analytic and metaphysical techniques of medieval natural philosophers. These developments did not emerge from a vacuum. Although they represent truly profound changes in the way science was done, they should be viewed as late developments in a process begun in the Middle Ages. Without the foundations laid down in the Middle Ages, and described in this study, seventeenth-century scientists could not have challenged prevailing opinions about the physical world, because there would have been little of substance to challenge in physics, astronomy, and cosmology.

From the three pre-conditions described earlier in this chapter, two vital and unique entities emerged during the late twelfth and the thirteenth centuries in Western Europe: the universities and the class of theologian-natural philosophers. Their existence permitted natural philosophy to develop in ways that had never before been feasible, and to flourish. The developments within natural philosophy laid the foundations for scientific inquiry that came to fruition in the seventeenth century.

## ON THE RELATIONSHIP BETWEEN MEDIEVAL AND EARLY MODERN SCIENCE

Despite the considerable achievements discussed in this book, the medieval period in Western Europe has been grossly underestimated, and even maligned, almost as if fate had chosen it as history's scapegoat. At least two momentous events of the seventeenth century were responsible for this state of affairs: the process of the Scientific Revolution itself and the condemnation of Galileo by the Catholic Church in 1633. In the seventeenth century's long struggle to repudiate Aristotelian natural philosophy, the latter was made to appear monolithic, rigid, unimaginative, and totally inadequate. Its supporters were presented as insensitive dullards, as "logic-choppers" and "word-mongers" who were opposed to progress. The image of Aristotelian natural philosophers and scholastics was one of an old guard dedicated to the status quo.

No one contributed more to the formation of this image than did Galileo, who, in his famous *Dialogue on the Two Chief World Systems* (1632), left an indelible picture of Scholastic ineptitude and wrongheadedness. It was in this work that Galileo created Simplicio, a fictional character named after Simplicius, the well-known sixth-century Greek commentator on the works of Aristotle. Although many Aristotelians accepted the telescope as a valid scientific instrument, Simplicio declares that "following in the footsteps of other Peripatetic philosophers of my group, I have considered as fallacies and deceptions of the lenses those things which other people have admired as stupendous achievements."[42] Galileo also emphasized Simplicio's slavish dependence on Aristotle. In denying that the earth is an orbiting planet, Simplicio appeals to Aristotle, who, as Simplicio explains, raised serious, though unresolved, objections to the earth's rotation. "And since he [Aristotle] raised the difficulty without solving it, it must," insists Simplicio, "necessarily be very difficult of solution, if not entirely impossible."[43] By his literary and artistic genius, Galileo fashioned a powerful caricature that was applied to all

Aristotelian natural philosophers, not only to those in the seventeenth century but also retroactively to those who lived in the Middle Ages.

Galileo's devastating critique was reinforced from many sources. Toward the end of the seventeenth century, the great English philosopher John Locke, who had been taught medieval Aristotelian philosophy at Oxford University, characterized scholasticism as little more than futile verbal gymnastics. In his *Essay Concerning Human Understanding* (1690), Book III, chapter 9, Locke characterized the schoolmen as "the great mintmasters" of empty terms. What passed for "subtlety" and acuteness among them was "nothing but a good expedient to cover their ignorance, with a curious and inexplicable web of perplexed words." By "unintelligible terms," they sought "to procure the admiration of others." Locke spoke for most of the nonscholastic scientists and philosophers of his century.

By such sustained criticism, Aristotelian natural philosophers were made to appear foolish, ineffectual, and irrelevant. What could this group possibly contribute to the new science? Nothing whatever, according to its critics. Indeed, real science could emerge only by the total rejection of Aristotelian natural philosophy and the works of its defenders. Because seventeenth-century Aristotelianism was not distinguished from it, medieval scholasticism was regarded as merely an earlier version of the same thing. Thus the whole of scholasticism – from its beginnings in the thirteenth century to its effective termination at the end of the seventeenth – stood condemned.

The condemnation of Galileo in 1633, for upholding the Copernican heliocentric system, aggravated the situation immeasurably, for the Church was viewed as defending and preserving Aristotelianism by coercion. Indeed, throughout its history Aristotelian natural philosophy was seen as the creature of a church ready to exterminate any scientific ideas that it viewed as potentially threatening.

Thus a completely false view of medieval natural philosophy has been perpetuated and the contextual and substantive medieval achievements in natural philosophy described in this volume have been ignored by those modern historians who have judged late medieval developments by the attitudes that emerged in the seventeenth century when the new science was struggling to overthrow the Aristotelian world view. The true nature of medieval contributions to the seventeenth century is far more accurately depicted in an analogy drawn from the late Middle Ages. In the late thirteenth century, in Italy, the course of the history of medicine was altered significantly when human dissection was allowed for postmortems. Shortly after, human dissection was introduced into medical schools, where it soon became institutionalized as part of the anatomical training of medical students. Human dissection had been forbidden in the ancient world, except in Egypt, where it was also banned by the second century A.D. It was never permitted in the Islamic world.

Hence, its introduction into the Latin West, made without serious objection from the Church, was a momentous occurrence. Dissection of cadavers continued to be used primarily in teaching, although irregularly, until the end of the fifteenth century. Rarely was it employed in research to advance scientific knowledge of the human body. The revival of human dissection and its incorporation into medical training throughout the Middle Ages were basic to the significant anatomical progress that was made by such keen anatomists as Leonardo da Vinci (1452–1519), Bartolomeo Eustachi (ca. 1500–1574), Andreas Vesalius (1514–1564), Gabriele Falloppio (1523–1562), and Marcello Malpighi (1628–1694).

What human dissection in the Middle Ages did for the study of anatomy in the sixteenth and seventeenth centuries, the translations, the universities, the theologian-natural philosophers, and the medieval version of Aristotelian natural philosophy did collectively for the Scientific Revolution in the seventeenth century. These vital features of medieval science prepared the way for the next eight hundred years of uninterrupted scientific development, development that began in Western Europe and spread around the world.

## ON THE RELATIONSHIP BETWEEN EARLY AND LATE MEDIEVAL SCIENCE

If the achievements of the late Middle Ages in natural philosophy and science bear such a positive relationship to the early modern period, is it not plausible to suppose that a similar relationship might obtain between the early Middle Ages, say, the period from approximately 500 to 1150, and the late Middle Ages? In a word, no. The relationships are radically different. With a few minor exceptions, Greek science was absent during the early Middle Ages. Euclidian geometry, for example, was virtually non-existent. The Greco-Arabic science that entered Western Europe in the twelfth century was not merely the enrichment of a somewhat less developed Latin science. It signified a dramatic break with the past and a new beginning. Logic, science, and natural philosophy were henceforth institutionalized in the newly developed universities. Early medieval authors, such as Macrobius, Martianus Capella, and Isidore of Seville, were still read in the late Middle Ages, but they were no longer weighty authorities. They had been displaced by Aristotle and a host of Greco-Arabic authors who would in turn be supplemented by indigenous scholars of the late Middle Ages.

## GRECO-ARABIC-LATIN SCIENCE: A TRIUMPH OF THREE CIVILIZATIONS

Although we have seen that science can be traced back to the ancient civilizations of Egypt and Mesopotamia, the modern science that emerged in the seventeenth century in Western Europe was the legacy of a scientific tradition that began in Ancient Greek and Hellenistic civilization, was further nurtured and advanced in the far-flung civilization of Islam, and was brought to fruition in the civilization of Western Europe, beginning in the late twelfth century. For this reason the science and natural philosophy that I have discussed in this volume may be appropriately described as "Greco-Arabic-Latin" science. My primary concern has been to explicate the last, or Latin, part of this extraordinary tripartite, sequential process. The collective achievement of these three civilizations, despite their significant linguistic, religious, and cultural differences, stands as one of the greatest examples of multiculturalism in recorded history. It is an example of multiculturalism in its best sense. It was possible only because scholars in one civilization recognized the need to learn from scholars of another civilization. Latin scholars in the twelfth century recognized that all civilizations were not equal. They were painfully aware that with respect to science and natural philosophy their civilization was manifestly inferior to that of Islam. They faced an obvious choice: learn from their superiors or remain forever inferior. They chose to learn and launched a massive effort to translate as many Arabic texts into Latin as was feasible. Had they assumed that all cultures were equal, or that theirs was superior, they would have had no reason to seek out Arabic learning, and the glorious scientific history that followed might not have occurred.

If we move back a few centuries, we see that the same phenomenon occurred in the civilization of Islam. By the late eighth century, Islamic scholars, speaking and writing in Arabic, became aware of a vast body of Greek scientific literature. Recognizing that the absence of scientific literature in the Arabic language was a serious cultural and intellectual deficiency, they set out to translate Greek science and natural philosophy into Arabic. If Muslim scholars had assumed that they had nothing to learn from "Dead Greek Pagans," and if Latin Christian scholars similarly had assumed that they had nothing to learn from "Dead Muslim Heretics," the great translation movements of the Middle Ages would never have occurred, impoverishing history and delaying scientific development by many centuries. Fortunately, this did not happen, and we can look back upon Greco-Arabic-Latin science and natural philosophy as one continuous development. It was truly a progression toward modern science and represents one of the most glorious chapters in human history.

# Notes

## PREFACE

1. See Koyré's *Études Galiléennes*. 3 fascicules. *I. A l'aube de la science classique* (Paris: Hermann, 1939), 9. For an excellent analysis of the various opinions and attitudes on this question, see John E. Murdoch, "Pierre Duhem and the History of Late Medieval Science and Philosophy in the Latin West," in R. Imbach and A. Maierù, eds., *Gli studi di filosofia medievale fra otto e novecento* (Rome: Edizioni di Storia e Letteratura, 1991), 253–302.
2. See my article "Medieval Science and Natural Philosophy," in James M. Powell, ed., *Medieval Studies, An Introduction* (Syracuse: Syracuse University Press, 1992), 369.

## CHAPTER 1

1. Peter Brown, *Power and Persuasion in Late Antiquity: Towards a Christian Empire* (Madison: University of Wisconsin Press, 1992), 122.
2. See D. S. Wallace-Hadrill, *The Greek Patristic View of Nature* (Manchester: Manchester University Press and New York: Barnes & Noble, 1968), 6.
3. *Saint Basil Exegetic Homilies*, translated by Sister Agnes Clare Way, vol. 46 of *The Fathers of the Church: A New Translation* (Washington, D.C.: Catholic University of America Press, 1963), 5.
4. Ibid., 17.
5. Ibid., 144.
6. Ibid., 72.
7. Quoted from Williston Walker, *A History of the Christian Church* (New York: Scribner's, 1949), 135.
8. James Westfall Thompson and Edgar Nathaniel Johnson, *An Introduction to Medieval Europe, 300–1500* (New York: W. W. Norton & Co., 1937), 645–646.
9. John L. LaMonte, *The World of the Middle Ages: A Reorientation of Medieval History* (New York: Appleton-Century-Crofts, 1949), 255.
10. I have slightly altered these quotations from the translation by Ernest Brehaut, *An Encyclopedist of the Dark Ages* (New York: Columbia University Press, 1912), 133.

CHAPTER 2

1. Translated in M. D. Chenu, *Nature, Man, and Society in the Twelfth Century: Essays on New Theological Perspectives in the Latin West*, selected, edited, and translated by Jerome Taylor and Lester K. Little (Chicago: University of Chicago Press, 1968; originally published in French in 1957), 10.
2. Translated in Chenu, ibid., 11. William made the remark in his *Philosophy of the World* (*Philosophia mundi*).
3. Chenu, ibid., n. 20.
4. Translated by Michael McVaugh in Edward Grant, *A Source Book in Medieval Science* (Cambridge, Mass.: Harvard University Press, 1974), 35.
5. Harry A. Wolfson, "Revised Plan for the Publication of a *Corpus Commentariorum Averrois in Aristotelem*," *Speculum* (1963): 88.

CHAPTER 3

1. Ibn Khaldun, *The Muqaddimah: An Introduction to History*, translated from the Arabic by Franz Rosenthal, 3 vols. (Princeton: Princeton University Press, 1958; corrected, 1967), 3: 117–118.

CHAPTER 4

1. Aristotle, *Physics* 4.4.212a.5–7, translated by R. P. Hardie and R. K. Gaye in *The Complete Works of Aristotle*, the revised Oxford translation edited by Jonathan Barnes (Princeton: Princeton University Press, 1984).
2. Aristotle, *Physics* 4.4.212b.12–14.
3. My translation from Thomas Aquinas's commentary on Aristotle's *Physics*. The translation was made from *S. Thomae Aquinatis In octo libros De physico auditu sive Physicorum Aristotelis commentaria*, new edition by Angeli-M. Pirotta (Naples: D'Auria Pontificius, 1953), 203 (bk. 4, lecture 7, par. 917).
4. Aristotle, *On the Heavens* 4.2.308b.19–20, translated by J. L. Stocks in *The Complete Works of Aristotle*.
5. In his *Physics*, bk. 8, ch. 6 (259a.29–31), Aristotle declared that "In the course of our argument directed to this end we established the fact that everything that is in motion is moved by something, and that the mover is either unmoved or in motion." Translated by R. P. Hardie and R. K. Gaye in *The Complete Works of Aristotle*.
6. D. R. Dicks, *Early Greek Astronomy to Aristotle* (Ithaca, N.Y.: Cornell University Press, 1970), 200–201.
7. Dante, *The Divine Comedy, Paradiso*, Canto XXXIII, translated by Laurence Binyon in *The Portable Dante*, edited by Paolo Milano (New York: Penguin Books, 1947), 544. In commenting on this line, Charles S. Singleton declared that "the last verse of the poem bears the image of Aristotle's unmoved mover, the spheres turning in desire of him, being moved by desire of him." See Dante Alighieri, *The Divine Comedy*, translated, with a commentary, by Charles S. Singleton, 3 vols. (Princeton: Princeton University Press, 1970–1975), 3: 590.

8. Cited in John Bartlett, *Familiar Quotations,* 16th edition, edited by Justin Kaplan (Boston: Little, Brown and Co., 1992), 786.

9. Ibid., 530.

10. Aristotle, *Physics* 2.2.194b.13–14, translated by R. P. Hardie and R. K. Gaye in *The Complete Works of Aristotle.*

11. Translated in David Knowles, *The Evolution of Medieval Thought* (Baltimore: Helicon Press, 1962), 200.

12. Ibid.

13. Etienne Gilson, *History of Christian Philosophy in the Middle Ages* (London: Sheed and Ward, 1955), 642, n.17. For a quite different attitude toward Aristotle, see my description of Albertus Magnus in chapter 7.

14. Dante, *Inferno* I.

CHAPTER 5

1. Aristotle, *On the Heavens,* bk. 2, ch.1, 283b.26–30, translated by W. K. C. Guthrie (Loeb Classical Library, 1960).

2. My translation from Boethius's *De aeternitate mundi* (*On the Eternity of the World*) in *Boethii Daci Opera, Topica-Opuscula,* vol. 6, pt. 2, edited by Nicolaus Georgius Green-Pedersen (Copenhagen, 1974), 355–357.

3. From Siger, *Question on the Eternity of the World,* in *St. Thomas Aquinas, Siger of Brabant, St. Bonaventure, On the Eternity of the World* (*De aeternitate mundi*) translated from the Latin with an introduction by Cyril Vollert, Lottie H. Kendzierski, and Paul M. Byrne (Milwaukee: Marquette University Press, 1964), 93.

4. Thomas Aquinas, *Summa Theologiae,* pt. 1, qu. 46, art. 1 in *St. Thomas Aquinas, Siger of Brabant, St. Bonaventure, On the Eternity of the World* (*De aeternitate mundi*), 66.

5. Translated in Edward Grant, *A Source Book in Medieval Science* (Cambridge, Mass.: Harvard University Press, 1974), 47.

6. Armand Maurer, "Boetius of Dacia and the Double Truth," *Mediaeval Studies* 17 (1955): 238. Maurer uses an alternative spelling for "Boethius."

7. All but article 139 are reproduced, with a few changes, from Grant, *Source Book,* 48–49. The translations were made from the Latin text edited by H. Denifle and E. Chatelain, *Chartularium Universitatis Parisiensis,* 4 vols. (Paris: Fratrum Delalain, 1889–1897), 1: 543–555.

8. Albert discussed this in his *Questions on Aristotle's Generation and Corruption,* bk. 2, qu. 14. See Edward Grant, "Science and Theology in the Middle Ages," in David C. Lindberg and Ronald L. Numbers, *God and Nature: Historical Essays on the Encounter between Christianity and Science* (Berkeley: University of California Press, 1986), 68.

9. My translation from Albert of Saxony, *Questiones De generatione et corruptione,* as cited in Anneliese Maier, *Metaphysische Hintergründe der spätscholastischen Naturphilosophie* (Rome, 1955), 39–40.

10. From Buridan's *Questions on the Eight Books of the Physics of Aristotle,* bk. 4, qu. 8, as translated in Grant, *Source Book,* 50–51.

11. Translation altered from Grant, *Source Book,* 50.

CHAPTER 6

1. Translated from Nicholas Bonetus, *Habes Nicholai Bonetti . . . quattuor volumina: Metaphysicam, videlicet naturalem phylosophiam, predicamenta, necnon theologiam naturalem . . .* (Venice, 1505), fol. 63, col. 2, by Edward Grant, in Grant, *Much Ado About Nothing: Theories of Space and Vacuum from the Middle Ages to the Scientific Revolution* (Cambridge: Cambridge University Press, 1981), 43.

2. Translated from Heytesbury's *Regule solvendi sophismata* (Venice, 1494), fol. 39, by Ernest Moody, which appears, slightly modified, in Marshall Clagett, *Science of Mechanics in the Middle Ages* (Madison: University of Wisconsin Press, 1959), 270.

3. From Campanus of Novara's *Theory of the Planets*, sec. IV, translated by Francis S. Benjamin, Jr., and G. J. Toomer, *Campanus of Novara and Medieval Planetary Theory, Theorica planetarum*, edited with an introduction, English translation, and commentary by Francis S. Benjamin, Jr., and G. J. Toomer (Madison: University of Wisconsin Press, 1971), 183.

4. My translation from Buridan, *Questions on De caelo*, bk. 2, qu. 22, in Ernest A. Moody, *Iohannis Buridani Quaestiones super libris quattuor De caelo et mundo* (Cambridge, Mass.: Mediaeval Academy of America, 1942), 230.

5. Nicole Oresme, *Le Livre du ciel et du monde*, bk. 2, ch. 25, in the edition and translation by Albert D. Menut and Alexander J. Denomy (Madison: University of Wisconsin Press, 1968), 535.

6. See Richard Sorabji, "Infinity and the Creation," in Richard Sorabji, ed., *Philoponus and the Rejection of Aristotelian Science* (Ithaca, N.Y.: Cornell University Press, 1987), 167.

7. From St. Bonaventure's *Commentary on the Sentences*, bk. 2, dist. 1, p. 1, a. 1, qu. 2, in *St. Thomas Aquinas, Siger of Brabant, St. Bonaventure, On the Eternity of the World (De aeternitate mundi)*, translated from the Latin with an introduction by Cyril Vollert, Lottie H. Kendzierski, and Paul M. Byrne (Milwaukee: Marquette University Press, 1964), 107.

8. From Aquinas's *Summa Theologiae*, pt. 1, qu. 46, art. 2, as translated in ibid., 66.

9. Thomas Aquinas, *On the Eternity of the World*, in ibid., 23.

10. Ibid., 20.

11. From Robert Holkot, *In quattuor libros Sententiarum quaestiones (Questions on the Four Books of the Sentences)* (Lyon, 1518; reprinted in facsimile Frankfurt: Minerva, 1967), bk. 2, qu. 2, sig. hii, recto, col. 2 [unfoliated]. The translation appears in Grant, *Much Ado About Nothing*, 351, n. 130.

12. Translated in Grant, *A Source Book in Medieval Science* (Cambridge, Mass.: Harvard University Press, 1974), 559, col. 1.

13. This is one of five corollaries in a chapter titled "That God is not mutable in any way." Translated in Grant, *Source Book*, 557, col. 1. The translation is repeated in Grant, *Much Ado About Nothing*, 135.

CHAPTER 7

1. John Buridan, *Questions on De caelo*, bk. 1, qu. 19. My translation from *Iohannis Buridani Quaestiones super libris quattuor De caelo et mundo*, edited by Ernest

A. Moody (Cambridge, Mass.: Mediaeval Academy of America, 1942), 87–90.

2. It is in Paris at the Bibliothèque Nationale (fonds latin, MS. 6752).

3. My translation from ibid., fol. 7v.

4. The questions cited in this section are drawn from "The Catalog of Questions on Medieval Cosmology, 1200–1687," in Edward Grant, *Planets, Stars, & Orbs: The Medieval Cosmos, 1200–1687* (Cambridge: Cambridge University Press, 1994), 681–746.

5. From Bonaventure's *Commentary on the Four Books of the "Sentences,"* bk. 2, dist. 14, pt. 2, art. 2, qu. 2, in his *Opera Omnia*, vol. 2 (1885). Translated in Grant, *Planets, Stars, & Orbs*, 575.

6. From Roger Bacon's *Opus Maius* vi. 1, translated by A. C. Crombie, in Crombie, *Robert Grosseteste and the Origins of Experimental Science, 1100–1700* (Oxford: Clarendon Press, 1953), 141.

7. *Questions on the Metaphysics*, bk. 2, qu. 1, in the reprint edition titled *Johannes Buridanus, Kommentar zur Aristotelischen Metaphysik* (Paris, 1588; reprinted Frankfurt: Minerva, 1964), fol. 8r, col. 1.

8. Ibid., folios 8v, col. 2–9r, col. 1.

9. Bert Hansen (ed.), *Nicole Oresme and the Marvels of Nature: A Study of His "De causis mirabilium" with Critical Edition, Translation, and Commentary* (Toronto: Pontifical Institute of Mediaeval Studies, 1985), 63.

10. Buridan, *Questions on the Metaphysics*, bk. 2, qu. 1, fol. 9r, col. 1.

11. Ibid., bk. 2, qu. 2, fol. 9v, col. 2. The translation is by Ernest A. Moody in "Buridan, Jean," *Dictionary of Scientific Biography*, edited by Charles C. Gillispie, 16 vols. (1970–1980), 2: 605.

12. My translation from Buridan's *Questions on De caelo*, bk. 2, qu. 9 on p. 164 of Moody's edition.

13. See his *Parts of Animals* 4.12.694a.15; 4.13.695b.19; and *Generation of Animals* 2.4.739b.20 and 2.5.744a.37.

14. *On the Heavens* 3.4.302b.21, translated by J. L. Stocks in *The Complete Works of Aristotle*, the revised Oxford translation edited by Jonathan Barnes (Princeton: Princeton University Press, 1984).

15. Translated by Julius R. Weinberg, in Weinberg, *A Short History of Medieval Philosophy* (Princeton: Princeton University Press, 1964), 239.

16. Ibid., 239, n. 3.

17. From Buridan's *Questions on De caelo*, bk. 2, qu. 22, as translated in Edward Grant, "Scientific Thought in Fourteenth-Century Paris: Jean Buridan and Nicole Oresme," in *Machaut's World: Science and Art in the Fourteenth Century*, edited by Madeleine P. Cosman and Bruce Chandler. *Annals of the New York Academy of Sciences* (New York: The New York Academy of Sciences, 1978), 314: 121, n. 29.

18. From Ockham's *Commentary on the Sentences*, bk. 1, distinction 14, qu. 2, G, translated by Weinberg, *A Short History of Philosophy*, 239.

19. Nicole Oresme, *Le Livre du ciel et du monde*, bk. 2, ch. 22, in the edition and translation by Albert D. Menut and Alexander J. Denomy (Madison: University of Wisconsin Press, 1968), 507.

20. What follows on Blasius of Parma is based on the discussion in Edward Grant, *Much Ado About Nothing: Theories of Space and Vacuum from the Middle*

*Ages to the Scientific Revolution* (Cambridge: Cambridge University Press, 1981), 89–92. For the full title of Blasius's work, see page 422.

21. In the title of an article, "The Analytic Character of Late Medieval Learning: Natural Philosophy without Nature," in *Approaches to Nature in the Middle Ages: Papers of the Tenth Annual Conference of the Center for Medieval & Early Renaissance Studies*, edited by Lawrence D. Roberts. *Medieval and Renaissance Texts and Studies* (Binghamton, N.Y.: Center for Medieval & Early Renaissance Studies, 1982), 16: 171–213.

22. Translated by Walter Ong, *Ramus, Method, and the Decay of Dialogue: From the Art of Discourse to the Art of Reason* (Cambridge, Mass.: Harvard University Press, 1958), 144.

23. The translated quotations from the Coimbra Jesuits in this paragraph appear in Grant, *Planets, Stars, & Orbs*, 110. The Latin texts appear in *Commentariorum Collegii Conimbricensis Societatis Iesu In octo libros Physicorum Aristotelis Stagiritae* (Lyon, 1602), pt. 1, bk. 3, ch. 8, qu. 2, art. 3, col. 544.

24. Translated by G. E. R. Lloyd in *Hippocratic Writings*, translated by J. Chadwick and W. N. Mann and edited by G. E. R. Lloyd (Penguin Books, 1978), 38.

25. 480b.26ff. Translated by G. R. T. Ross in *The Complete Works of Aristotle*.

26. Translated by Nancy Siraisi from Peter of Abano's *Conciliator differentiarum philosophorum et praecipue medicorum*, in Nancy Siraisi, *Medieval and Early Renaissance Medicine: An Introduction to Knowledge and Practice* (Chicago: University of Chicago Press, 1990), 67.

27. Ibid., 68.

28. From Blasius's *Questions on the Physics*, bk. 1, qu. 3. Translated in Brian Lawn, *The Rise and Decline of the Scholastic 'Quaestio Disputata' with Special Emphasis on Its Use in the Teaching of Medicine and Science* (Leiden: E. J. Brill, 1993), 59.

29. See Sarah Waterlow, *Nature, Change, and Agency in Aristotle's "Physics": A Philosophical Study* (Oxford: Clarendon Press, 1982), 33–34. For a summary, see David C. Lindberg, *The Beginnings of Western Science: The European Scientific Tradition in Philosophical, Religious, and Institutional Context, 600 B.C. to A.D. 1450* (Chicago: University of Chicago Press, 1992), 52–53.

30. Cited by Lynn White, Jr., "Cultural Climates and Technological Advance in the Middle Ages," reprinted from *Viator* 2 (1971), in Lynn White, Jr., *Medieval Religion and Technology: Collected Essays* (Berkeley and Los Angeles: University of California Press, 1978), 221.

31. This remark about Bernard of Chartres was reported by John of Salisbury, in his *Metalogicon*, iii. 4. It was translated by D. D. McGarry in McGarry's *The Metalogicon of John of Salisbury* (Berkeley: University of California Press, 1962), 167.

32. Translated in A. C. Crombie, *Styles of Scientific Thinking in the European Tradition*, 3 vols. (London: Duckworth, 1994), 1:25.

33. Translated in Edward A. Synan, "Albertus Magnus and the Sciences," in *Albertus Magnus and the Sciences: Commemorative Essays 1980*, edited by James A. Weisheipl (Toronto: The Pontifical Institute of Mediaeval Studies, 1980), 11. The translation was made from Albert's commentary on Aristotle's *De anima (On the Soul)*, tract 2, ch. 3.

34. Ibid. Translated from Albertus's *Commentary on the Physics*, bk. 8, tract 1, ch.

14. Albertus's attitude toward Aristotle stands in striking contrast to that of Averroes, as described earlier in chapter 4.

35. See Buridan, *Questions on De caelo*, bk. 1, qu. 18, in Moody's edition, 86.

36. *Questions on the Physics*, bk. 8, qu. 12, translated from the edition of Paris, 1509, in Grant, *Source Book*, 275.

37. Charles Coulston Gillispie, *The Edge of Objectivity: An Essay in the History of Scientific Ideas* (Princeton: Princeton University Press, 1969), 11.

## CHAPTER 8

1. See the following two books by William Newman: *The Summa perfectionis of Pseudo-Geber: A Critical Edition, Translation and Study* (Leiden and New York: E. J. Brill, 1991) and *Gehennical Fire: The Lives of George Starkey, an American Alchemist in the Scientific Revolution* (Cambridge, Mass.: Harvard University Press, 1994).

2. Translated by Edward A. Synan, "Albertus Magnus and the Sciences," in *Albertus Magnus and the Sciences: Commemorative Essays 1980*, edited by James A. Weisheipl (Toronto: Pontifical Institute of Mediaeval Studies, 1980), 9–10.

3. A. I. Sabra, "The Appropriation and Subsequent Naturalization of Greek Science in Medieval Islam: A Preliminary Statement," *History of Science* 25 (1987): 225–226.

4. A. I. Sabra, "Science and Philosophy in Medieval Islamic Theology," *Zeitschrift für Geschichte der Arabisch-Islamischen Wissenschaften*, vol. 9 (Frankfurt: Institut für Geschichte der Arabisch-Islamischen Wissenschaften an der Johann Wolfgang Goethe-Universität, 1994), 5.

5. F. E. Peters, *Aristotle and the Arabs: The Aristotelian Tradition in Islam* (New York: New York University Press; London: University of London Press, 1968), 220.

6. Translated in M. Montgomery Watt, *The Faith and Practice of al-Ghazali* (London: George Allen and Unwin, 1953), 33.

7. Ibid.

8. Ibid., 34.

9. Ibid., 37.

10. From *Al-Ghazali's Tahafut al-Falasifah [Incoherence of the Philosophers]*, translated into English by Sabih Ahmad Kamali (Pakistan Philosophical Congress Publication, no. 3, 1963), 249.

11. Ibid., 3.

12. Ibn Khaldun, *The Muqaddimah: An Introduction to History*, translated from the Arabic by Franz Rosenthal, 3 vols. (Princeton: Princeton University Press, 1958; corrected, 1967), vol. 3, ch. 6, sec. 30, 246–258.

13. Ibid., 250.

14. Ibid., 251–252.

15. Ibid., 257–258.

16. Averroes, *On the Harmony of Religion and Philosophy: A translation with introduction and notes, of Ibn Rushd's Kitab fasl al-maqal, with its appendix (Damima) and an extract from Kitab al-kashf 'an manahij al-adilla*, by George F. Hourani (London: Luzac, 1976), 44.

17. From J. M. Hussey, *The Byzantine World*, 2nd edition (London: Hutchinson University Library, 1961), 148.
18. Deno J. Geanakoplos, *Medieval Western Civilization and the Byzantine and Islamic Worlds: Interactions of Three Cultures* (Lexington, Mass.: D. C. Heath and Co., 1979), 145.
19. Steven Runciman, *The Last Byzantine Renaissance* (Cambridge: Cambridge University Press, 1970), 1.
20. Ibid., 97.
21. Quoted by Runciman, ibid., 94.
22. Donald M. Nicol, *Church and Society in the Last Centuries of Byzantium* (Cambridge: Cambridge University Press, 1979), 47–48.
23. Nicol, ibid., 50.
24. Norman H. Baynes, *Byzantine Studies and Other Essays* (London, 1955; reprinted 1960), 72.
25. Quoted in Stanley Guralnick, *Science and the Ante-Bellum American College* (Philadelphia: American Philosophical Society, 1975), Memoirs, vol. 109, 60. I am grateful to my colleague, Professor James Capshew, for the reference.
26. Guralnick, ibid., 61.
27. From Buridan's *Questions on the Metaphysics*, bk. 2, qu. 1, as translated by Ernest A. Moody in "Buridan, Jean," *Dictionary of Scientific Biography*, edited by Charles C. Gillispie, 16 vols. (1970–1980), 2:605.
28. Johannes Kepler, *New Astronomy*, translated by William H. Donahue (Cambridge: Cambridge University Press, 1992), 51. Kepler's *New Astronomy* (*Astronomia Nova*) was first published in 1609.
29. John Locke, *An Essay Concerning Human Understanding*, bk. 2, ch. 13, par. 22, in *The Philosophical Works of John Locke*, edited by J. A. St. John, 2 vols. (London, 1903–1905), 1:295.
30. Translated from Gassendi's *Syntagma philosophicum* in his *Opera Omnia*, vol. 1 (Lyon, 1658; reprinted by Friedrich Frommann Verlag, 1964), 183, col. 1.
31. Ibid., 182, col. 2, in *The Selected Works of Pierre Gassendi*, edited and translated by Craig B. Brush (New York: Johnson Reprint, 1972), 386. Also cited in Edward Grant, *Much Ado About Nothing:Theories of Space and Vacuum from the Middle Ages to the Scientific Revolution* (Cambridge: Cambridge University Press, 1981), 390, n. 169.
32. From Hobbes's *De corpore* (*Concerning Body*), translated from Latin into English in *The English Works of Thomas Hobbes of Malmesbury*, edited by William Molesworth, 16 vols. (London, 1839–1845), 1:93. Also cited in Grant, *Much Ado About Nothing*, 390, n. 169.
33. From Gassendi's *Syntagma philosophicum*, as translated in *The Concepts of Space and Time, Their Structure and Their Development*, edited by Milič Čapek (Dordrecht and Boston: D. Reidel, 1976), 93.
34. See H. G. Alexander, ed., *The Leibniz–Clarke Correspondence, Together with Extracts from Newton's "Principia" and "Opticks"* (Manchester: University of Manchester Press, 1956), 32.
35. Quoted from I. Bernard Cohen and Richard S. Westfall, eds., *Newton: Texts, Backgrounds, Commentaries* (New York: W. W. Norton & Co., 1995), 233. I have added the bracketed phrase on the basis of a note to the first law by Cohen and Westfall.

36. See Synan, "Albertus Magnus and the Sciences," in Weisheipl, *Albertus Magnus and the Sciences*, 10.

37. Translated by Mary Martin McLaughlin, *Intellectual Freedom and Its Limitations in the University of Paris in the Thirteenth and Fourteenth Centuries* (New York: Arno Press, 1977), 96.

38. From Oresme's *Le Livre du ciel et du monde*, bk. 1, ch. 29, in the edition and translation by Albert D. Menut and Alexander J. Denomy (Madison: University of Wisconsin Press, 1968), 197.

39. Translated from Buridan's *Questions on De caelo*, bk. 2, qu. 14, in *Iohannis Buridani Quaestiones super libris quattuor De caelo et mundo*, edited by Ernest A. Moody (Cambridge, Mass.: Mediaeval Academy of America, 1942), 184.

40. Translated from Buridan, *Questions on De caelo*, bk. 2, qu. 17, in Moody, 208.

41. Peter Dear, *Mersenne and the Learning of the Schools* (Ithaca, N.Y.: Cornell University Press, 1988), 1.

42. See *Galileo Galilei: Dialogue Concerning the Two Chief World Systems – Ptolemaic & Copernican*, translated by Stillman Drake, foreword by Albert Einstein (Berkeley and Los Angeles: University of California Press, 1962), 336.

43. Ibid., 379.

# Bibliography

The bibliography is in two parts. The first part is an essay in which works are grouped and related thematically. The second part lists the relevant books and articles in alphabetical order by author. The citations in the first part are as brief as possible, containing only the author, or occasionally, the journal, and date of publication enclosed by brackets, as, for example, [Stahl, 1962]. For the full citation, the reader must look under the author's name in the second part. Where there is more than one entry for an author, they are listed chronologically, earliest to most recent.

Because there is an enormous literature on the many aspects of medieval science and natural philosophy, I have confined this bibliography largely to books and articles directly relevant to the themes and topics considered in this study.

## PART I. BIBLIOGRAPHICAL ESSAY

### GENERAL WORKS ON THE HISTORY OF MEDIEVAL SCIENCE

Two works on medieval history will provide a larger perspective against which to view this book. One, Thompson and Johnson, 1937, is an old-fashioned, detailed factual introduction containing an enormous amount of information and genealogical tables; the other, Cantor, 1993, is an imaginative and exciting interpretation of medieval life in most of its manifestations.

Certain encyclopedic treatises and bibliographies provide useful information on medieval science and technology. Foremost among these is Gillispie, 1970–1980. A perusal of the index in volume 16, for example, reveals articles on approximately four hundred authors from the ancient and medieval periods prior to 1480. Sarton, 1927–1948, serves as a supplement for authors up to 1400, for the numerous authors omitted from Gillispie. Additional articles on major medieval scientists and natural philosophers appear in Strayer, 1982–1989 (see the comprehensive index in volume 13). For a helpful list of 1,470 articles and books, see Kren, 1985.

The most important source of bibliographical data for medieval science and technology and for the history of science as a whole is the "current bibliographies" (formerly known as "critical bibliographies") published annually in *Isis*, the journal of the History of Science Society. These have been reorganized and republished: see *Isis*, 1971–1982, 1980, 1990, 1989. Current bibliographies continue to appear annually.

Surveys of medieval science as a whole are hardly numerous. By far the best and most comprehensive is the latest, Lindberg, 1992, an extensive description and analysis of ancient and medieval science. Some years earlier, Lindberg edited a collection of articles contributed by leading scholars in the field: Lindberg, 1978. A brief work with an accompanying bibliographical essay is Grant, 1971. Still useful and interesting are two older accounts by Crombie, 1952, and Dijksterhuis, 1961, 99–219 and 248–253, the latter emphasizing medieval intellectual currents and physical science. In a brief volume, Weisheipl, 1959, conveys a sense of medieval scholastic science. Devoted almost exclusively to magic and the pseudo-sciences, but including valuable biographical and bibliographical information, is the encyclopedic work by Thorndike, 1923–1958.

For primary sources in English translation ranging over the whole of medieval science, see Grant, 1974. Dales, 1973, presents a mix of primary sources and modern scholarly assessments. In an unusual and useful volume intended "primarily as an anthology of the kinds of pictorial and diagrammatic materials to be found in the scientific literature of antiquity and the Middle Ages" (p. x), Murdoch, 1984, has assembled hundreds of illustrations from manuscript sources and provided them with extensive and informative captions.

Collections of articles by modern historians of medieval science and natural philosophy have been published in recent years, often by way of reprinting articles from a variety of journals, some now difficult to obtain. Here is a list of some of these collections, alphabetized by author: Clagett, 1979; Courtenay, 1984; Crombie, 1991; Eastwood, 1989; Goldstein, 1985; Grant, 1981; Kibre, 1984; King, 1993; Lindberg, 1983; Moody, 1975; North, 1989; Rosenthal, 1990; Sabra, 1994; and Weisheipl, 1985.

### CHAPTER 1. THE ROMAN EMPIRE AND THE FIRST SIX CENTURIES OF CHRISTIANITY

For good general summaries of the Roman Empire, see Charlesworth, 1968, and Wells, 1984; for its end, see Kagan, 1992. With an emphasis on the Eastern Roman Empire, Brown, 1992, describes the way in which Christians and Christianity entered into the power structure of the Roman Empire. An overall view of science in the Roman Empire period appears in Clagett, 1957, 99–156. In recent years, our understanding of the late ancient Greek commentators on Aristotle, especially the sixth-century authors Simplicius and John Philoponus, has been greatly enlarged with studies and translations. Of importance is a series of articles in Sorabji, 1987. Among a number of translations, I mention two of works by Philoponus: Wildberg, 1987, 1988.

For a list of works on the Latin encyclopedic, or handbook, tradition of science, see Kren, 1985, 15–21. A description of that tradition for the early Middle Ages appears in Stahl, 1962. Articles on each of the seven liberal arts, which were associated with the encyclopedic tradition, appear in Wagner, 1983. Translations from significant treatises in this tradition are found in Stahl, 1952; Brehaut, 1912; and Stahl, Johnson, Burge, 1971, 1977, vol. 2, which also contains a translation of Martianus Capella's *Marriage of Philology and Mercury*, a treatise on the seven liberal arts.

Christian attitudes toward pagan, secular learning in late antiquity and the early Middle Ages in both East and West form an important aspect of the overall development of Western science. The attitudes of Latin authors are described in Ellspermann, 1949. Opinions of the Church fathers and others on a wide range of natural phenomena and scientific questions are contained in commentaries on the creation of the world as described in Genesis. For discussions, see Wallace-Hadrill, 1968; Duhem, 1913–1959, vol. 2, 393–504; Robbins, 1912; and Thorndike, 1923–1958, vol. 1. Saint Basil's famous essay on how Christians should use pagan literature appears in Wilson, 1975.

## CHAPTER 2. THE NEW BEGINNING

By the twelfth century, even before the translations, Europe was in the process of a radical transformation. Indeed some have called it a "Renaissance." The most famous book advocating this interpretation is Haskins, 1957 (*Renaissance*), which also contains chapters on the revival of science and philosophy in the twelfth century. For others in this genre, see Paré, Brunet, Tremblay, 1933, and Young, 1969, which not only contain material concerning the controversy over designating the twelfth century as a renaissance but also compare it specifically to the Italian Renaissance; see also Hollister, 1969.

Although the science of the early Middle Ages was modest, it improved in the eleventh century and quite dramatically in the twelfth century. Stiefel, 1985, describes an increased emphasis on rational investigation into natural phenomena. Adelard of Bath's varied activities led Cochrane, 1994, to call him, with considerable justice, "The First English Scientist." An excellent example of the twelfth century's new, more critical approach toward science is Adelard of Bath's *Natural Questions (Quaestiones Naturales)*, which simultaneously exhibited scorn for the older Latin learning. Adelard's treatise is translated in Gollancz, 1920. For William of Conches's role, see Cadden, 1995. John of Salisbury's defense of the old ways has been translated in McGarry, 1955. In two essays, "Nature and Man – The Renaissance of the Twelfth Century" and "The Platonisms of the Twelfth Century," Chenu, 1968, emphasizes the importance of the microcosm-macrocosm analogy, Platonism, and an increased naturalism in the twelfth century. Thorndike, 1923–1958, vol. 2, chs. 36, 37, 39, describes occult aspects of the thought of Adelard of Bath, William of Conches, and Bernard Silvester.

An excellent overall summary of medieval translating activity covering the tenth to thirteenth centuries appears in Lindberg, 1978 ("Transmission"). An equally splendid presentation of translating activity, focused on medieval Spain, appears in Burnett, 1992. Lists of translations for the two most significant translators of the Middle Ages, Gerard of Cremona, who translated from Arabic to Latin, and William of Moerbeke, who translated from Greek to Latin, appear in Grant, 1974, 35–38 (Michael McVaugh's list for Gerard) and 39–41 (for Moerbeke's translations). Crombie, 1959 (*Medieval*), vol. 1, 37–47, provides valuable tables for the translations that include author, work, translator, language of original translation, and place and date of translation. For translations into Latin from Greek, see Muckle, 1942, 1943.

For fairly detailed discussions of major Greek and Arabic scientific authors – for example, Aristotle, Euclid, Galen, Ptolemy, Avicenna, Averroes, al-Razi, Alhazen (Ibn al-Haytham) – and the transmission of their works and ideas to the medieval Latin West, see under their respective names in Gillispie, 1970–1980. Of these important scientists and natural philosophers, Aristotle is undoubtedly the most significant for medieval thought. With the inclusion of tables, Dod, 1982, presents an excellent summary of the translators, translations, and dissemination of Aristotle's works.

## CHAPTER 3. THE MEDIEVAL UNIVERSITY

Education for most of the early Middle Ages is covered in Riché, 1976. On cathedral schools, the precursors to the universities, see Southern, 1982; Gabriel, 1969; and Williams, 1954, 1964.

A large body of modern literature now exists on all aspects of university life. Among general works, especially noteworthy is the most recent, Ridder-Symoens, 1992, which examines the medieval university from every significant standpoint (structures, management and resources, students, curriculum, and faculties). An earlier wide-ranging collection of some thirty articles appears in Ijsewijn and Paquet, 1978. Cobban, 1975, provides a good general account of the universities. An exciting introduction to medieval universities, with its fascinating marginalia of Latin terms and phrases, is Piltz, 1981. Wieruszowski, 1966, is a solid, informative survey.

Of older works, the classic, and still useful, study is the revised edition of Rashdall, 1936. Haskins, 1957 (*Rise*), though brief, remains perhaps the most readable account. Also useful is Daly, 1961. An excellent and still unsurpassed collection of source readings in translation is Thorndike, 1944. Two outstanding, specialized studies (on the nations and scholarly privileges) are Kibre, 1948, 1962.

Because the universities of Paris, Oxford, and Bologna were the prototypes of all other European universities and were also the most significant centers of scientific thought, books and articles describing their histories are of special importance. For Paris, see Leff, 1968; Glorieux, 1933–1934, 1965–1966, 1971; Ferruolo, 1985; and Halphen, 1949. In a significant study, McLaughlin, 1977, describes the struggle for academic freedom at Paris (also see McLaughlin, 1955, and Courtenay, 1989). Articles touching upon important themes at the University of Paris are McKeon, 1964, Post, 1929, 1934, and Courtenay, 1988. Among a number of histories devoted to Oxford University, or in which that university plays a significant part, see Leff, 1968, and Cobban, 1988. A list of masters and scholars to 1500 appears in Emden, 1957. Articles on a variety of interesting and important themes are Callus, 1943; Lytle, 1978, 1984; and Weisheipl, 1964, 1966. For a general history of the University of Bologna, see Sorbelli, 1940. The intellectual and social climate are described in Zaccagnini, 1926. The larger relations of the university are described in Hyde, 1972; on disputations, see Matsen, 1977.

Works describing and analyzing medieval science and natural philosophy in a university context are Grant, 1984; Beaujouan, 1963; and Bullough, 1966, which contains summaries of the early histories of the medical schools of Salerno, Montpellier, Bologna, and Paris, with a few references to Padua; Lemay, 1976; Weisheipl, 1969; Courtenay, 1980; Sylla, 1985; and Kibre and Siraisi, 1978. An unusual book related to learning and the universities is Murray, 1978; see especially Part II, Arithmetic, 141–210, and Part III, Reading and Writing, 213–257, which deal with university life and the intellectual elite.

CHAPTER 4. WHAT THE MIDDLE AGES INHERITED
FROM ARISTOTLE

The standard English translations of Aristotle's works are the Oxford translation, originally completed in 1931 and recently issued as Aristotle, 1984. A convenient, but selective, edition is Aristotle, 1941, which includes the complete *Physics, On Generation and Corruption, Metaphysics, On the Soul,* and the *Short Physical Treatises,* with most of *On the Heavens,* and only brief extracts from the biological works. Lloyd, 1968, presents a lucid introduction to Aristotle's physical and biological thought and provides an excellent bibliography of additional readings, 316–317. A good guide to the difficult concepts in Aristotle's *Physics* is Waterlow, 1982. Also useful is Ross, 1949. For a briefer account, see Lindberg, 1992, 46–68 ("Aristotle's Philosophy of Nature").

CHAPTER 5. THE RECEPTION AND IMPACT OF ARISTOTELIAN
LEARNING AND THE REACTION OF THE CHURCH
AND ITS THEOLOGIANS

In this volume, we have seen that Aristotle exercised a more profound and extensive influence on medieval science and natural philosophy than any other Greco-Arabic author. Lemay, 1962, describes the influx of Aristotelian thought even before the major translations of his works were underway. By the 1220s, the works of Aristotle were known in Latin translation. Their reception, however, proved difficult and at times tumultuous. The introduction and study of Aristotle's scientific treatises are described at some length by Duhem, 1913–1959, vol. 5, especially chs. 8–13, 233–580; Steenberghen, 1970; and Callus, 1943.

The reaction to, and consequences of, Aristotelian philosophy are described in a number of histories of medieval philosophy: Gilson, 1955; Copleston, 1957, 1953; and Knowles, 1962; in a brief, but brilliant, account, Weinberg, 1965 opposes Gilson's interpretation, arguing that the philosophical achievements of the fourteenth-century nominalists exceeded that of their thirteenth-century predecessors. Articles concerned with the problems of reception and reaction are Lohr, 1982, and Grant, 1985.

As the culminating reaction to the issues raised by the works of Aristotle, the Condemnation of 1277 has attracted considerable attention. A first condemnation, in 1270, is treated in Wéber, 1970, and Wippel, 1977. The apprehension of the Church and its theologians concerning the adverse influence of Arabic and Greek science and philosophy can be gauged from the list of philosophers' errors compiled between 1270 and 1274 by Giles of Rome (Aegidius Romanus) in his *Errores philosophorum* – Koch, 1944. Hissette, 1977, presents the Latin texts of all 219 propositions and also attempts to trace their sources and give their significance. L. Fortin and Peter D. O'Neill have translated all the articles into English in Fortin and O'Neill, 1963. Condemned articles related to medieval science and natural philosophy have been translated in Grant, 1974, 45–50. On the role played by the condemnation, see Bianchi, 1990; for its influence on natural philosophy and cosmology, see Grant, 1979.

## CHAPTER 6. WHAT THE MIDDLE AGES DID WITH ITS ARISTOTELIAN LEGACY

### The terrestrial region

Interest in problems of motion and change were pervasive and ubiquitous in the late Middle Ages. The best overall, extended treatment of medieval kinematics and dynamics is Clagett, 1959. Excellent summary accounts of medieval concepts of motion are Murdoch and Sylla, 1978, and Maier, 1982. Good briefer descriptions appear in Lindberg, 1992, 290–307, and Weisheipl, 1982. Two fundamental treatises on proportionality theory, with its concept of "ratios of ratios" and the latter's application to laws of motion, have been edited and translated: see Crosby, 1955, and Grant, 1966. Selections from these two treatises appear in Grant, 1974, 158–159, 306–312 (Oresme), 292–305 (Bradwardine). On proportion and proportionality theory, see Murdoch, 1963.

A special center of activity was Merton College, Oxford, where there was intense interest in the quantification of qualities. On this theme, see Sylla, 1971, 1987. For more on the Merton school, or Oxford calculators, see Coleman, 1975, and Sylla, 1988, who describes the fortunes of the Oxford tradition. Nicole Oresme's treatise on the quantification of qualities, perhaps the most important in the Middle Ages, has been edited and translated in Clagett, 1968.

There is a considerable literature on medieval dynamics, or the causes of motion. For numerous selections, including Bradwardine's dynamic law of motion, impetus theory, and the free fall of bodies, see Clagett, 1959, Part III, 421–625. On impetus theory, see Maier, 1982 ("Significance"), 76–102; Franklin, 1976; and Wallace, 1981 ("Galileo"). Ernest Moody has studied medieval concepts of motion with an eye to their possible influence on Galileo. In his most significant article, Moody argues – Moody, 1951 – that Galileo's early dynamics (the Pisan period) was derived ultimately from the discussion of Avempace (Ibn Bajja), a twelfth-century Muslim, who lived in Spain. During his subsequent Paduan period, Moody shows – Moody, 1966 – that Galileo adopted Buridan's ideas of permanent impetus and his explanation of free fall, as well as accepting the Mertonian analysis of accelerated motion. For another account of the relationship of scholastic impetus theory and Galileo's ideas of the causes of motion, see Maier, 1982 ("Galileo"), 103–123.

On the kinematics and dynamics of motion in void and plenum, and the concepts of internal resistance and mixed bodies, see Part I of Grant, 1981 (*Much Ado*), and also see Grant's bibliography for other relevant works. Numerous source selections on medieval kinematics, dynamics, and the concept of vacuum as well as the possibility of motion in it, appear in Grant, 1974, 234–253 (kinematics), 253–312 (dynamics), and 324–360 (vacuum).

### The celestial region

An informative summary of medieval astronomy is Pedersen, 1978. Pedersen also describes the widely used astronomical text, *Theorica planetarum (Theory of the Planets)*, in Pedersen, 1981, and he also translates the *Theorica* in Grant, 1974, 451–465. Sharing popularity with the *Theorica planetarum* was John of Sacrobosco's *Tractatus de spera (Treatise on the Sphere)*, which has been edited and translated by Thorndike, 1949. Using the title *Theorica planetarum*, Campanus of Novara wrote a much more sophisticated Ptolemaic treatise, which has been published by Benjamin and Toomer, 1971.

Duhem, 1913–1959, was the first scholar to study medieval cosmology broadly and in depth. For a recent extensive study of medieval cosmology with a lengthy bibliography, see Grant, 1994. On the eternity of the world, see Dales, 1990, and, for the Latin texts, Dales and Agerami, 1991. Other articles on the eternity of the world are Wissink, 1990; Wippel, 1981; and Kretzmann, 1985, published 1988. Whether the celestial bodies were thought to be alive or not is described by Dales, 1980. A discussion of celestial movers appears in Weisheipl, 1961, and Wolfson, 1973. Celestial influences on the terrestrial region are described in North, 1986, 1987, and Grant, 1994, 569–617. For a view of medieval cosmology through the writings of individual scholastics, see Steneck, 1976, and Litt, 1963. There is much valuable detail and discussion on cosmology, astrology, and astronomy in North, 1988. Readings on a number of cosmological themes appear in Grant, 1974, 494–568.

## CHAPTER 7. MEDIEVAL NATURAL PHILOSOPHY, ARISTOTELIANS, AND ARISTOTELIANISM

For an analysis of late medieval natural philosophy, see Wallace, 1988; Maier, 1982 ("Achievements"); Murdoch, 1982; and North, 1992. The raw materials of natural philosophy, the questions (*questiones*), are studied in Bazàn, Wippel, Fransen, and Jacquart, 1985; Wippel, 1982; and Lawn, 1963, 1993. For a history of medieval Aristotelianism, see Steenberghen, 1970, and for Oxford alone, Callus, 1943. Lohr, 1988, cites the literature on medieval scholastic Aristotelian commentators, and the articles and books about them from approximately 1967 to 1987.

Speculation on why Aristotelianism endured in Europe for approximately five centuries is offered in Grant, 1978. He considers other aspects of Aristotelianism in Grant, 1987, with its self-explanatory title. For a different emphasis, see Thijssen, 1991. For the fortunes of medieval scholastic natural philosophy in the Renaissance, see Schmitt, 1983, who describes the many facets and complexities of early modern Aristotelianism; also see Iorio, 1991.

A small body of literature exists on the interrelationship of natural philosophy and other disciplines: for its relationship to music, see Murdoch, 1976; for logic Murdoch, 1989; for theology Sylla, 1975, and Grant, 1986; and for medicine Siraisi, 1990, ch. 3, especially 65–67.

### CHAPTER 8. HOW THE FOUNDATIONS OF EARLY MODERN SCIENCE WERE LAID IN THE MIDDLE AGES

Because chapter 8 is based upon earlier chapters, the bibliography will be confined to those topics that are first discussed in this final chapter.

A good, general history of the Arabs is Dunlop, 1971. For brief surveys of Islamic science, see Arnaldez and Massignon, 1963; Sabra, 1976, 1982–1989; and Kennedy, 1970. Brief descriptions of individual sciences are found in Goldstein, 1986, and Saliba, 1982, 1994 (astronomy); Rashed, 1994, and Berggren, 1986 (mathematics); and Meyerhof, 1931, and Dunlop, 1971, 204–250 (medicine). Perceptive, but quite different, discussions about the nature and development of Islamic science are given by Sabra, 1987, and Huff, 1993.

A general introduction to Islamic philosophy is Leaman, 1985. For the translations of Aristotle's works into Arabic, see Peters, 1968 (*Aristoteles*). On Aristotle's natural philosophy in Islam, see Peters, 1968, (*Aristotle*), and Lettinck, 1994; a good summary chapter appears in Dunlop, 1971, 172–203. On the *kalam*, see Wolfson, 1976, and Sabra, 1994 ("Science"). Physical theories in which Greek natural philosophy and Islamic theology, physics, and cosmology were interrelated within the *kalam* are the subject of Dhanani, 1994. On the relations of religion with natural philosophy and science, see Watt, 1953; Kamali, 1963; Marmura, 1975; and Hourani, 1976; on the relations between philosophy and theology, see Guillaume, 1931; Watt, 1962; Kamali, 1963; and Hourani, 1976.

On learning, and attitudes toward learning, in the Byzantine Empire, see Geanakoplos, 1979; Hussey, 1961; Runciman, 1970; and Nicol, 1979. A summary description of Byzantine science appears in Théodoridès, 1963.

The influence of late medieval science and natural philosophy on the Scientific Revolution of the seventeenth century has been frequently discussed. Are medieval science and the new science of the sixteenth and seventeenth centuries continuous or discontinuous? The most recent, balanced, interpretation, which includes a brief history of the problem, is Lindberg, 1992, 355–368 ("The Legacy of Ancient and Medieval Science"). Lindberg argues that there is discontinuity when the "global aspects" (p. 367) of science in the seventeenth century are emphasized, but that continuities are manifest when the focus is on specific disciplines.

Among those who have spoken for discontinuity, the best-known argument is that of Koyré, 1939, 1943, 1956. For an illuminating description and analysis of Koyré's approach to the history of science, see Murdoch, 1987. Also supportive of the discontinuity argument is McMullin, 1965, 1968. Rosen, 1961, argues that recent scholarship has upheld Burckhardt's judgment that modern science began in the Renaissance when medieval scholastic attitudes were largely abandoned. Among historians of medieval science, Murdoch, 1991, opposes the continuity thesis and expresses his sentiments in a splendid interpretation of the role Pierre Duhem played in shaping the modern approach to late medieval science. Murdoch argues that "seen properly and completely, what one was about in the fourteenth century not only did not lead to Galileo and the like, but was not even pointing in that direction" (p. 279). For Murdoch's earlier expressions of this point of view, see Murdoch, 1974, and Murdoch and Sylla, 1978, 250.

On the side of scientific continuity stands Pierre Duhem, who, in all the works we have cited thus far, was convinced of a direct medieval influence on the achievements of the Scientific Revolution of the seventeenth century. Those who support Duhem's general claim frequently impose significant qualifications. Two authors who maintain that seventeenth-century science inherited a well-formulated and essential methdology from the Middle Ages are John Herman Randall, Jr., and Alistair C. Crombie. From the former, see Randall, 1940. In a reprint of his article, Randall added Latin texts and emphasized the link to modern science by changing the title Randall, 1961. For Crombie, see Crombie, 1953, 290–319 and Crombie, 1959. Anneliese Maier's position is presented in Maier, 1982 ("Achievements"). In Moody, 1966, the very title of the article reveals the author's strong conviction that there was continuity between Galileo and his fourteenth-century predecessors, especially John Buridan. Two other more extensive studies – Wallace, 1981 (*Prelude*), and Lewis, 1980 – have sought to "fill in," to use Murdoch's phrase, the gaps in the sixteenth century between Galileo and his medieval predecessors and thus to reveal genuine continuity.

## PART II. BIBLIOGRAPHY

Aristotle. *The Basic Works of Aristotle.* Edited with an introduction by Richard McKeon. New York: Random House, 1941.
   *The Complete Works of Aristotle: The Revised Oxford Translation.* Edited by Jonathan Barnes. 2 vols. Princeton: Princeton University Press, 1984.
Arnaldez, R., and L. Massignon. "Arabic Science." In *History of Science: Ancient and Medieval Science from the Beginnings to 1450.* Edited with a general preface by René Taton, and translated [from French] by A. J. Pomerans. New York: Basic Books, 1963, 385–421.
Bazàn, B. C., J. F. Wippel, G. Fransen, and D. Jacquart. *Les Questions disputée et les questions quodlibétiques dans les facultés de théologie, de droit et de médicine.* Typologie des sources du Moyen Age occidental, edited by L. Genicot, fasc. 44–5. Turnhout, 1985.
Beaujouan, Guy. "Motives and Opportunities for Science in the Medieval Universities." In A. C. Crombie, ed., *Scientific Change.* New York: Basic Books, 1963, 219–236.
Benjamin, Francis, Jr., and G. J. Toomer, eds. and trans. *Campanus of Novara and Medieval Planetary Theory, "Theorica planetarum."* Madison: University of Wisconsin Press, 1971.
Berggren, J. L. *Episodes in the Mathematics of Medieval Islam.* New York: Springer-Verlag, 1986.
Bianchi, Luca. *Il vescovo e i filosofi: La condanna Parigina del 1277 e l'evoluzione dell'Aristotelismo scolastico.* Bergamo: Pierluigi Lubrina Editore, 1990.
Brehaut, Ernest, trans. *An Encyclopedist of the Dark Ages: Isidore of Seville.* New York: Columbia University Press, 1912.
Brown, Peter. *Power and Persuasion in Late Antiquity: Towards a Christian Empire.* Madison: University of Wisconsin Press, 1992.
Bullough, Vern L. *The Development of Medicine as a Profession: The Contribution of*

*the Medieval University to Modern Medicine*. New York: Hafner Publishing Co., 1966.

Burnett, Charles. "The Translating Activity in Medieval Spain." In Salma Khadra Jayyusi, ed., *The Legacy of Muslim Spain*. Leiden: E. J. Brill, 1992. 1036–1058.

Cadden, Joan. "Science and Rhetoric in the Middle Ages: The Natural Philosophy of William of Conches." *Journal of the History of Ideas* 56 (1) (January 1995): 1–24.

Callus, D. A. "Introduction of Aristotelian Learning to Oxford." *Proceedings of the British Academy* 29 (1943): 229–281.

Cantor, Norman. *The Civilization of the Middle Ages. A Completely Revised and Expanded Edition of "Medieval History: The Life and Death of a Civilization."* New York: HarperCollins Publishers, 1993.

Charlesworth, Martin P. *The Roman Empire*. New York: Oxford University Press, 1968.

Chenu, M. D. *Nature, Man, and Society in the Twelfth Century: Essays on New Theological Perspectives in the Latin West*. Selected, edited, and translated by Jerome Taylor and Lester K. Little. Chicago: University of Chicago Press, 1968. Originally published in 1957.

Clagett, Marshall. *Greek Science in Antiquity*. London: Abelard-Schuman, 1957.

*The Science of Mechanics in the Middle Ages*. Madison: University of Wisconsin Press, 1959.

*Studies in Medieval Physics and Mathematics*. London: Variorum Reprints, 1979.

ed. and trans. *Nicole Oresme and the Medieval Geometry of Qualities and Motions: A Treatise on the Uniformity and Difformity of Intensities Known as "Tractatus de configurationibus qualitatum et motuum."* Edited with an introduction, English translation, and commentary by Marshall Clagett. Madison: University of Wisconsin Press, 1968.

Cobban, Alan B. *The Medieval Universities: Their Development and Organization*. London: Methuen & Co., 1975.

*The Medieval English Universities: Oxford and Cambridge to c. 1500*. Berkeley and Los Angeles: University of California Press, 1988.

Cochrane, Louise. *Adelard of Bath: The First English Scientist*. London: British Museum Press, 1994.

Coleman, Janet. "Jean de Ripa O.F.M. and the Oxford Calculators." *Mediaeval Studies* 37 (1975): 130–189.

Copleston, Frederick J. *A History of Philosophy*. Vol. 2, *Augustine to Scotus*; and Vol. 3, *Ockham to Suarez*. Westminster, Md.: Newman Press, 1957 (first published 1950), 1953.

Courtenay, William J. "The Effect of the Black Death on English Higher Education." *Speculum* 55 (1980): 696–714.

*Covenant and Causality in Medieval Thought: Studies in Philosophy, Theology, and Economic Practice*. London: Variorum Reprints, 1984.

"Teaching Careers at the University of Paris in the Thirteenth and Fourteenth Centuries." In A. L. Gabriel and P. E. Beichner, eds., *Texts and Studies in the History of Mediaeval Education*. No.18. U.S. Subcommission for the History of Universities. Notre Dame, Ind.: University of Notre Dame, 1988. (38 pp.).

"Inquiry and Inquisition: Academic Freedom in Medieval Universities." *Church History* 58 (1989): 168–181.

Crombie, A. C. *Robert Grosseteste and the Origins of Experimental Science, 1100–1700*. Oxford: Clarendon Press, 1953.

*Medieval and Early Modern Science*. 2 vols. New York: Doubleday, 1959. First printed as *Augustine to Galileo*. London, 1952.

"The Significance of Medieval Discussions of Scientific Method for the Scientific Revolution." In Marshall Clagett, ed., *Critical Problems in the History of Science*. Madison: University of Wisconsin Press, 1959. 79–101. (See also the comments on Crombie's paper by I. E. Drabkin, 142–147, and Ernest Nagel, 153–154.)

*Science, Optics, and Music in Medieval and Early Modern Thought*. London: Hambledon Press, 1991.

Crosby, H. Lamar, Jr., ed. and trans. *Thomas of Bradwardine: His Tractatus de proportionibus, Its Significance for the Development of Mathematical Physics*. Madison: University of Wisconsin Press, 1955.

Dales, Richard C. *The Scientific Achievement of the Middle Ages*. Philadelphia: University of Pennsylvania Press, 1973.

"The De-animation of the Heavens in the Middle Ages." *Journal of the History of Ideas* 41 (1980): 531–550.

*Medieval Discussions of the Eternity of the World*. Leiden: Brill, 1990.

Dales, Richard C., and Omar Agerami, eds. *Medieval Latin Texts on the Eternity of the World*. Leiden: Brill, 1991.

Daly, Lowrie J. *The Medieval University, 1200–1400*. New York: Sheed and Ward, 1961.

Dhanani, Alnoor. *The Physical Theory of Kalam: Atoms, Space, and Void in Basrian Mu'tazili Cosmology*. Islamic Philosophy, Theology, and Science: Texts and Studies, 14. Leiden: Brill, 1994.

Dijksterhuis. E. J. *The Mechanization of the World Picture*. Translated by C. Dikshoorn. Oxford: Clarendon Press, 1961. First published in Dutch in 1950.

Dod, Bernard G. "Aristoteles Latinus." In Norman Kretzmann, Anthony Kenny, and Jan Pinborg, eds., *The Cambridge History of Later Medieval Philosophy*. Cambridge: Cambridge University Press, 1982. 45–79.

Duhem, Pierre. *Le Système du monde, histoire des doctrines cosmologiques de Platon à Copernic*. 10 vols. Paris: Hermann, 1913–1959.

Dunlop, D. M. *Arab Civilization to A. D. 1500*. London: Longman, 1971.

Eastwood, Bruce. *Astronomy and Optics from Pliny to Descartes: Texts, Diagrams, and Conceptual Structures*. London: Variorum Reprints, 1989.

Ellspermann, Gerard L. *The Attitude of the Early Christian Latin Writer toward Pagan Literature and Learning*. Catholic University of America, Patristic Studies, 82. Washington, D.C., 1949.

Emden, A. B. *A Biographical Register of the University of Oxford to A.D. 1500*. 3 vols. Oxford: Clarendon Press, 1957.

Ferruolo, Stephen C. *The Origins of the University: The Schools of Paris and Their Critics, 1100–1215*. Stanford, Calif.: Stanford University Press, 1985.

Fortin, L., and Peter D. O'Neill. "Condemnation of 219 Propositions." In Ralph Lerner and Muhsin Mahdi, eds., *Medieval Political Philosophy: A Sourcebook*.

New York: Free Press of Glencoe, 1963. 337–354. Reprinted in Arthur Hyman and James J.Walsh, *Philosophy in the Middle Ages*. Indianapolis: Hackett Publishing Co., 1973. 540–549.

Franklin, Allan. *The Principle of Inertia in the Middle Ages*. Boulder, Colo.: Associated University Press, 1976.

Gabriel, Astrik L. "The Cathedral Schools of Notre-Dame and the Beginning of the University of Paris." In Astrik L. Gabriel, *Garlandia; Studies in the History of the Mediaeval University*. Notre Dame, Ind.: Mediaeval Institute, University of Notre Dame; and Frankfurt: J. Knecht, 1969. 39–64.

Geanakoplos, Deno J. *Medieval Western Civilization and the Byzantine and Islamic Worlds: Interactions of Three Cultures*. Lexington, Mass.: D. C. Heath and Co., 1979.

Gillispie, Charles C., ed. *Dictionary of Scientific Biography*. 16 vols. New York: Charles Scribner's Sons, 1970–1980.

Gilson, Etienne. *History of Christian Philosophy in the Middle Ages*. London, 1955.

Glorieux, Palémon. *Répertoire des maîtres en théologie de Paris au xiiie siècle*. 2 vols. Paris: J. Vrin, 1933–1934.

*Aux origines de la Sorbonne*. 2 vols. Paris: J. Vrin, 1965–1966.

*La Faculté des arts et ses maîtres au xiiie siècle*. Paris: J. Vrin, 1971.

Goldstein, Bernard R. *Theory and Observation in Ancient and Medieval Astronomy*. London: Variorum Reprints, 1985.

"The Making of Astronomy in Early Islam." *Nuncius* 1 (1986): 79–92.

Gollancz, Hermann. *Dodi Venechdi (Uncle and Nephew), the Work of Berachya Hanakdan, now edited from the MSS. at Munich and Oxford, an English Translation, Introduction, etc., to Which Is Added the First English Translation from the Latin of Adelard of Bath's Quaestiones Naturales*. London, 1920.

Grant, Edward. *Physical Science in the Middle Ages*. New York: John Wiley & Sons, 1971. Reprinted by Cambridge University Press, 1977.

*A Source Book in Medieval Science*. Cambridge, Mass.: Harvard University Press, 1974.

"Aristotelianism and the Longevity of the Medieval World View." *History of Science* 16 (1978): 93–106.

"The Condemnation of 1277, God's Absolute Power, and Physical Thought in the Late Middle Ages." *Viator* 10 (1979): 211–244.

*Much Ado About Nothing: Theories of Space and Vacuum from the Middle Ages to the Scientific Revolution*. Cambridge: Cambridge University Press, 1981.

*Studies in Medieval Science and Natural Philosophy*. London: Variorum Reprints, 1981.

"Science and the Medieval University." In James M. Kittelson and Pamela J. Transue, eds., *Rebirth, Reform and Resilience: Universities in Transition, 1300–1700*. Columbus: Ohio State University, 1984. 68–102.

"Issues in Natural Philosophy at Paris in the Late Thirteenth Century." *Medievalia et Humanistica*, new series, no.13 (1985): 75–94.

"Science and Theology in the Middle Ages." In David C. Lindberg and Ronald L. Numbers, eds., *God and Nature: Historical Essays on the Encounter between Christianity and Science*. Berkeley and Los Angeles: University of California Press, 1986. 49–75.

"Ways to Interpret the Terms 'Aristotelian' and 'Aristotelianism' in Medieval and Renaissance Natural Philosophy." *History of Science* 25 (1987): 335–358.

*Planets, Stars, & Orbs: The Medieval Cosmos, 1200–1687*. Cambridge: Cambridge University Press, 1994.

ed. and trans. *Nicole Oresme, "De proportionibus proportionum" and "Ad pauca respicientes."* Madison: University of Wisconsin Press, 1966.

Guillaume, Alfred. "Philosophy and Theology." In *The Legacy of Islam*. Edited by the late Sir Thomas Arnold and Alfred Guillaume. Oxford: Oxford University Press, 1931. 239–283.

Halphen, Louis, et al. *Aspects de l'Université de Paris*. Paris, 1949.

Haskins, Charles H. "The Life of Mediaeval Students as Illustrated by Their Letters." In Charles H. Haskins, *Studies in Mediaeval Culture*. Oxford: Clarendon Press, 1929. 1–35.

*The Rise of Universities*. Ithaca, N.Y.: Great Seal Books, Cornell University Press, 1957. First published by Brown University, 1923.

*The Renaissance of the 12th Century*. Cleveland: World Publishing Co. [Meridian Books], 1957. First published in 1927.

Hissette, Roland. *Enquête sur les articles condamnés à Paris le 7 Mars 1277*. Louvain: Publications Universitaires; and Paris: Vander-Oyez, S.A., 1977.

Hollister, Warren C., ed. *The Twelfth Century Renaissance*. New York, 1969.

Hourani, George F., trans. *On the Harmony of Religion and Philosophy. A Translation with introduction and notes, of Ibn Rushd's Kitab fasl al-maqal, with its appendix (Damima) and an extract from Kitab al-kashf 'an manahij al-adilla*. London: Luzac, 1976.

Huff, Toby E. *The Rise of Early Modern Science: Islam, China, and the West*. Cambridge: Cambridge University Press, 1993.

Hussey, J. M. *The Byzantine World*. 2nd ed. London: Hutchinson University Library, 1961.

Hyde, J. K. "Commune, University, and Society in Early Medieval Bologna." In John W. Baldwin and Richard A. Goldthwaite, eds., *Universities in Politics: Case Studies from the Late Middle Ages and Early Modern Period*. Baltimore: Johns Hopkins University Press, 1972. 17–46.

Ijsewijn, Josef, and Jacques Paquet, eds. *The Universities in the Late Middle Ages. Mediaevalia Lovaniensia*. Series 1, studia 6. Leuven: Leuven University Press, 1978.

Iorio, Dominick A. *The Aristotelians of Renaissance Italy: A Philosophical Exposition*. Lewiston, N.Y.: Edwin Mellen Press, 1991.

*Isis. Cumulative Bibliography, 1913–1965*. 5 vols. Edited by Magda Whitrow. London: Mansell, 1971–1982.

*Cumulative Bibliography, 1966–1975*. Edited by John Neu. London: Mansell, 1980.

*Cumulative Bibliography, 1976–1985*. Edited by John Neu. Vol. 1. London: Mansell, 1990. Vol. 2. Boston: G. K. Hall & Co., 1989.

Kagan, Donald, ed. *The End of the Roman Empire: Decline or Transformation?* Third edition. Lexington, Mass.: D. C. Heath and Company, 1992.

Kamali, Sabih Ahmad. *Al-Ghazali's Tahafut al-Falasifah [Incoherence of the Philosophers]*. Translated into English by Sabih Ahmad Kamali. Pakistan Philosophical Congress, Publication, no. 3, 1963.

Kennedy, E. S. "The Arabic Heritage in the Exact Sciences." *Al-Abhath* 23 (1970): 327–344.

Kibre, Pearl. *The Nations in the Medieval Universities.* Cambridge, Mass.: Mediaeval Academy of America, 1948.

*Scholarly Privileges in the Middle Ages.* Cambridge, Mass.: Mediaeval Academy of America, 1962.

*Studies in Medieval Science: Alchemy, Astrology, Mathematics, and Medicine.* London: Hambledon Press, 1984.

Kibre, Pearl, and Nancy G. Siraisi. "The Institutional Setting: The Universities." In Lindberg, *Science in the Middle Ages,* 1978. 120–144.

King, David A. *Astronomy in the Service of Islam.* Variorum Collected Studies Series, 416. Aldershot, Eng.: Variorum Reprints, 1993.

Knowles, David. *The Evolution of Medieval Thought.* Baltimore: Helicon Press, 1962.

Koch, Josef, ed. *Giles of Rome, Errores philosophorum.* Translated by John O. Riedl. Milwaukee: Marquette University Press, 1944.

Koyré, Alexandre. *Études Galiléennes.* 3 fascicules. Paris: Hermann, 1939.

"Galileo and Plato." *Journal of the History of Ideas* 4 (1943): 400–428. Reprinted in Philip P. Wiener and Aaron Noland, eds., *Roots of Scientific Thought.* New York: Basic Books, 1957. 147–175.

"Les origines de la science moderne." *Diogène* 26 (1956): 14–42.

Kren, Claudia. *Medieval Science and Technology: A Selected, Annotated Bibliography.* New York: Garland Publishing, 1985.

Kretzmann, Norman. "Ockham and the Creation of the Beginningless World." *Franciscan Studies* 45 (1985; published 1988): 1–31.

Kretzmann, Norman, Anthony Kenny, and Jan Pinborg, eds. *The Cambridge History of Later Medieval Philosophy. From the Rediscovery of Aristotle to the Disintegration of Scholasticism, 1100–1600.* Cambridge: Cambridge University Press, 1982.

Lawn, Brian. *The Salernitan Questions: An Introduction to the History of Medieval and Renaissance Problem Literature.* Oxford: Clarendon Press, 1963.

*The Rise and Decline of the Scholastic "quaestio disputata," with special emphasis on its use in the teaching of medicine and science.* Education and Society in the Middle Ages and Renaissance, 2. Leiden: Brill, 1993.

Leaman, Oliver. *An Introduction to Islamic Philosophy.* Cambridge: Cambridge University Press, 1985.

Leff, Gordon. *Paris and Oxford Universities in the Thirteenth and Fourteenth Centuries: An Institutional and Intellectual History.* New York: John Wiley & Sons, 1968.

Lemay, Richard. *Abu Ma'shar and Latin Aristotelianism in the Twelfth Century: The Recovery of Aristotle's Natural Philosophy through Arabic Astrology.* Oriental Series, no. 38. Beirut: American University of Beirut, 1962.

"The Teaching of Astronomy in Medieval Universities, Principally at Paris in the 14th Century." *Manuscripta* 20 (1976): 197–217.

Lettinck, P. *Aristotle's "Physics" and Its Reception in the Arabic World. With an Edition of the Unpublished Parts of Ibn Bajja's "Commentary on the Physics."* Aristoteles Semitico-Latinus, 7. Leiden: Brill, 1994.

Lewis, Christopher. *The Merton Tradition and Kinematics in Late Sixteenth and Early Seventeenth Century Italy.* Padua: Antenore, 1980.

Lindberg, David C. "The Transmission of Greek and Arabic Learning to the West." In Lindberg, *Science in the Middle Ages*, 1978. 52–90.

*Studies in the History of Medieval Optics*. London: Variorum Reprints, 1983.

*The Beginnings of Western Science: The European Scientific Tradition in Philosophical, Religious, and Institutional Context, 600 B.C. to A.D. 1450*. Chicago: University of Chicago Press, 1992.

ed. *Science in the Middle Ages*. Chicago: University of Chicago Press, 1978.

Litt, Thomas. *Les corps célestes dans l'univers de Saint Thomas d'Aquin. Philosophes médiévaux*, Vol. 7. Louvain: Nauwelaerts, 1963.

Lloyd, G. E. R. *Aristotle: The Growth and Structure of His Thought*. Cambridge: Cambridge University Press, 1968.

Lohr, Charles H. "The Medieval Interpretation of Aristotle." In Kretzmann, Kenny, and Pinborg, *The Cambridge History of Later Medieval Philosophy*, 1982. 80–98

*Commentateurs d'Aristote au moyen-âge Latin, Bibliographie de la littérature secondaire récente; Medieval Latin Commentators, A Bibliography of Recent Secondary Literature*. Fribourg: Éditions Universitaires; Paris: Éditions du Cerf, 1988.

Lytle, Guy F. "The Social Origins of Oxford Students in the Late Middle Ages: New College c.1380–c.1510." In Ijsewijn and Paquet, *The Universities in the Late Middle Ages*, 1978. 426–454.

"The Careers of Oxford Students in the Later Middle Ages." In J. M. Kittelson and P. J. Transue, eds., *Rebirth, Reform and Resilience: Universities in Transition, 1300–1700*. Columbus: Ohio State University Press, 1984. 213–253.

McGarry, Daniel D. *The Metalogicon of John of Salisbury: a Twelfth-Century Defense of the Verbal and Logical Arts of the Trivium*. Translated with an introduction and notes by Daniel D. McGarry. Berkeley: University of California Press, 1962.

McKeon, P. R. "The Status of the University of Paris as *Parens scientiarum*." *Speculum* 39 (1964): 651–675.

McLaughlin, Mary Martin. "Paris Masters of the 13th and 14th Centuries and Ideas of Intellectual Freedom." *Church History* 24 (1955): 195–211.

*Intellectual Freedom and Its Limitations in the University of Paris in the Thirteenth and Fourteenth Centuries*. New York: Arno Press, 1977. (Ph.D. diss., Columbia University, 1952.)

McMullin, Ernan. "Medieval and Modern Science: Continuity or Discontinuity?" *International Philosophical Quarterly* 5 (1965): 103–129.

"Empiricism and the Scientific Revolution." In Charles S. Singleton, ed., *Art, Science, and History in the Renaissance*. Baltimore: Johns Hopkins University Press, 1968. 331–369.

Maier, Anneliese. "The Achievements of Late Scholastic Philosophy," 143–170; "Causes, Forces, and Resistance," 40–60; "Galileo and the Scholastic Theory of Impetus," 103–123; "The Nature of Motion," 21–39; and "The Significance of the Theory of Impetus for Scholastic Natural Philosophy," 76–102. In *On the Threshold of Exact Science, Selected Writings of Anneliese Maier on Late Medieval Natural Philosophy*. Edited and translated with an introduction by Steven D. Sargent. Philadelphia: University of Pennsylvania Press, 1982.

Marmura, Michael E. "Ghazali's Attitude to the Secular Sciences and Logic." In

George F. Hourani, ed., *Essays on Islamic Philosophy and Science*. Albany: State University of New York Press, 1975. 100–111.

Matsen, Herbert S. "Students 'Arts' Disputations at Bologna Around 1500, Illustrated from the Career of Alessandro Achillini (1463–1512)." *History of Education* 6, pt. 3 (1977): 169–181.

Meyerhof, Max. "Science and Medicine." In *The Legacy of Islam*. Edited by the late Sir Thomas Arnold and Alfred Guillaume. Oxford: Oxford University Press, 1931. 311–355.

Moody, Ernest A. "Galileo and Avempace: The Dynamics of the Leaning Tower Experiment." *Journal of the History of Ideas* 12 (1951): 163–193, 375–422.

"Galileo and His Precursors." In Carlo L. Golino, ed., *Galileo Reappraised*. Berkeley: University of California Press, 1966. 23–43.

*Studies in Medieval Philosophy, Science, and Logic: Collected Papers, 1933–1969*. Berkeley: University of California Press, 1975.

Muckle, J. T. "Greek Works Translated Directly into Latin before 1350." *Mediaeval Studies* 4 (1942): 33–42, and 5 (1943): 102–114.

Murdoch, John E. "The Medieval Language of Proportions: Elements of the Interaction with Greek Foundations and the Development of New Mathematical Techniques." In A. C. Crombie, ed., *Scientific Change*. New York: Basic Books, 1963. 237–271.

"Philosophy and the Enterprise of Science in the Later Middle Ages." In Y. Elkana, ed., *The Interaction Between Science and Philosophy*. Atlantic Heights, N.J.: Humanities Press, 1974. 55–57 (and the discussion on 104–105).

"Music and Natural Philosophy: Hitherto Unnoticed Questiones by Blasius of Parma(?)." In *Science, Medicine and the University: 1200–1500. Essays in Honor of Pearl Kibre, Part I*. Special editors Nancy G. Siraisi and Luke Demaitre. *Manuscripta* 20 (1976): 119–136.

"The Analytic Character of Late Medieval Learning: Natural Philosophy Without Nature." In L. D. Roberts, ed., *Approaches to Nature in the Middle Ages*. Binghamton, N.Y., 1982. 171–213. See also "Comment" by Norman Kretzmann, 214–220.

*Album of Science: Antiquity and the Middle Ages*. New York: Charles Scribner's Sons, 1984.

"Alexandre Koyré and the History of Science in America: Some Doctrinal and Personal Reflections." *History and Technology* 4 (1987): 71–79.

"The Involvement of Logic in Late Medieval Natural Philosophy." In Stefano Caroti, ed., *Studies in Medieval Natural Philosophy*. Firenze: Olschki, 1989. 3–28.

"Pierre Duhem and the History of Late Medieval Science and Philosophy in the Latin West." *Gli Studi di Filosofia Medievale fra Otto e Novecento*. Rome: Edizioni di Storia e Letteratura, 1991. 253–302.

Murdoch, John E., and Edith D. Sylla. "The Science of Motion." In Lindberg, *Science in the Middle Ages*, 1978. 206–264.

Murray, Alexander. *Reason and Society in the Middle Ages*. Oxford: Clarendon Press, 1978.

Nicol, Donald M. *Church and Society in the Last Centuries of Byzantium*. Cambridge: Cambridge University Press, 1979.

North, John D. "Celestial Influence – The Major Premiss of Astrology." In Paola

Zambelli, ed., *"Astrologi hallucinati"* Stars and the End of the World in Luther's Time. Berlin and New York: Walter de Gruyter, 1986. 45–100.

"Medieval Concepts of Celestial Influence, A Survey." In Patrick Curry, ed., *Astrology, Science and Society: Historical Essays.* Woodbridge, Suffolk: Boydell Press, 1987. 5–17.

*Chaucer's Universe.* Oxford: Clarendon Press, 1988.

*The Universal Frame: Historical Essays in Astronomy, Natural Philosophy, and Scientific Method.* London: Ronceverte, 1989.

"Natural Philosophy in Late Medieval Oxford." In J. I. Catto and Ralph Evans, eds., *The History of the University of Oxford.* Vol. 2. Oxford: Clarendon Press, 1992. Ch. 3, 65–102.

Paré, G., A. Brunet, and P. Tremblay. *La Renaissance du xii^e siècle: les écoles et l'enseignement.* Paris: J. Vrin; and Ottawa: Inst. d'études médiévales, 1933.

Pedersen, Olaf. "Astronomy." In Lindberg, *Science in the Middle Ages,* 1978. 303–337.

"The Origins of the Theorica planetarum." *Journal of the History of Astronomy* 12 (1981): 113–123.

Peters, F. E. *Aristoteles Arabus: The Oriental Translations and Commentaries on the Aristotelian "corpus."* Leiden: Brill, 1968.

*Aristotle and the Arabs: The Aristotelian Tradition in Islam.* New York: New York University Press, 1968.

Piltz, Anders. *The World of Medieval Learning.* Translated into English by David Jones. Totowa, N.J.: Barnes & Noble Books, 1981. First published in Sweden, 1978.

Post, Gaines. "Alexander III, the 'licentia docendi,' and the Rise of the Universities." In *C. H. Haskins Anniversary Essays.* Boston, 1929. 255–277.

"Parisian Masters as a Corporation, 1220–1246." *Speculum* 9 (1934): 421–445.

Randall, John Herman, Jr. "The Development of Scientific Method in the School of Padua." *Journal of the History of Ideas* 1 (1940): 177–206.

*The School of Padua and the Emergence of Modern Science.* Padua: Editrice Antenore, 1961.

Rashdall, Hastings. *The Universities of Europe in the Middle Ages.* Edited by F. M. Powicke and A. B. Emden. 3 vols. Oxford: Oxford University Press, 1936. Reprint, Oxford, 1988.

Rashed, Roshdi. *The Development of Arabic Mathematics: Between Arithmetic and Algebra.* Translated by A. F. W. Armstrong. Boston Studies in the Philosophy of Science, 156. Dordrecht: Kluwer Academic, 1994.

Riché, Pierre. *Education and Culture in the Barbarian West, Sixth Through Eighth Centuries.* Translated by J. Contreni and foreword by Richard E. Sullivan. Columbia: South Carolina University Press, 1976.

Ridder-Symoens, Hilde de. *A History of the University in Europe.* Vol.1: *Universities in the Middle Ages.* Cambridge: Cambridge University Press, 1992.

Robbins, F. E. *The Hexameral Literature: A Study of the Greek and Latin Commentaries on Genesis.* Chicago, 1912.

Rosen, Edward. "Renaissance Science as Seen by Burckhardt and His Successors." In Tinsley Helton, ed., *The Renaissance: A Reconstruction of the Theories and Interpretations of the Age.* Madison: University of Wisconsin Press, 1961. 77–103.

Rosenthal, Franz. *Science and Medicine in Islam: A Collection of Essays*. Aldershot, Eng.: Variorum Reprints, 1990.

Ross, W. D. *Aristotle*. 5th ed. rev. London: Methuen & Co., 1949.

Runciman, Steven. *The Last Byzantine Renaissance*. Cambridge: Cambridge University Press, 1970.

Sabra, A. I. "The Scientific Enterprise." In Bernard Lewis, ed., *Islam and the Arab World*. New York: Knopf, 1976. 181–192.

"Islamic Science." In Strayer, ed., *Dictionary of the Middle Ages*. 13 vols. 1982–1989. Vol. 11, 81–89.

"The Appropriation and Subsequent Naturalization of Greek Science in Medieval Islam: A Preliminary Statement." *History of Science* 25, pt. 3 (1987): 223–243.

*Optics, Astronomy, and Logic: Studies in Arabic Science and Philosophy* Aldershot, Hants, and Brookfield, Vt.: Variorum, 1994.

"Science and Philosophy in Medieval Islamic Theology." *Zeitschriftür für Geschichte der Arabisch-Islamischen Wissenschaften*, vol. 9. Frankfurt: Institut für Geschichte der Arabisch-Islamischen Wissenschaften an der Johann Wolfgang Goethe-Universität (1994): 1–42.

Saliba, George. "The Development of Astronomy in Medieval Islamic Society." *Arabic Studies Quarterly* 4 (1982): 211–225.

*A History of Arabic Astronomy: Planetary Theories During the Golden Age of Islam*. New York University Studies in Near Eastern Civilization. New York: New York University Press, 1994.

Sarton, George. *Introduction to the History of Science*. 3 vols. in 5 pts. Baltimore: Published for the Carnegie Institution of Washington by Williams & Wilkins Co., 1927–1948.

Schmitt, Charles B. *Aristotle and the Renaissance*. Cambridge, Mass.: Published for Oberlin College by Harvard University Press, 1983.

Siraisi, Nancy. *Medieval and Early Renaissance Medicine: An Introduction to Knowledge and Practice*. Chicago: University of Chicago Press, 1990.

Sorabji, Richard, ed. *Philoponus and the Rejection of Aristotelian Science*. Ithaca, N.Y.: Cornell University Press, 1987.

Sorbelli, A. *Storia della universita di Bologna*. Vol.1: *Il medio evo, saec. XI-XV*. Bologna, 1940.

Southern, R. W. "The Schools of Paris and the School of Chartres." In R. L. Benson and G. Constable, with C. D. Lanham, eds., *Renaissance and Renewal in the Twelfth Century*. Oxford: Oxford University Press, 1982. 113–137.

Stahl, William. *Roman Science: Origins, Development, and Influence to the Later Middle Ages*. Madison: University of Wisconsin Press, 1962.

trans. *Macrobius, Commentary on the Dream of Scipio*. New York: Columbia University Press, 1952.

Stahl, William, Richard Johnson, and E. L. Burge. *Martianus Capella and the Seven Liberal Arts*. 2 vols. New York: Columbia University Press, 1971, 1977.

Steenberghen, Fernand Van. *Aristotle in the West: The Origins of Latin Aristotelianism*. 2nd ed. Louvain: Nauwelaerts, 1970.

Steneck, Nicholas H. *Science and Creation in the Middle Ages: Henry of Langenstein (d.1397) on Genesis*. Notre Dame, Ind.: University of Notre Dame Press, 1976.

Stiefel, Tina. *The Intellectual Revolution in Twelfth-Century Europe.* New York: St. Martin's Press, 1985.

Strayer, Joseph, ed. *Dictionary of the Middle Ages.* 13 vols. New York: Charles Scribner's Sons, 1982–1989.

Sylla, Edith D. "Medieval Quantifications of Qualities: The 'Merton School.'" *Archive for History of Exact Sciences* 8 (1971): 9–39.

——— "Autonomous and Handmaiden Science: St. Thomas Aquinas and William of Ockham on the Physics of the Eucharist." In John Emery Murdoch and Edith Dudley Sylla, eds., *The Cultural Context of Medieval Learning.* Dordrecht and Boston: D. Reidel Publishing Co., 1975. 349–396.

——— "Science for Undergraduates in Medieval Universities." In Pamela O. Long, ed., *Science and Technology in Medieval Society.* New York: New York Academy of Sciences, 1985. 171–186.

——— "Mathematical Physics and Imagination in the Work of the Oxford Calculators: Roger Swineshead's On Natural Motions." In Edward Grant and John E. Murdoch, eds., *Mathematics and Its Applications to Science and Natural Philosophy in the Middle Ages: Essays in Honor of Marshall Clagett.* Cambridge: Cambridge University Press, 1987. 69–101.

——— "The Fate of the Oxford Calculatory Tradition." In Christian Wenin, ed., *L'homme et son univers au Moyen Age: Actes du 7e Congrès International de Philosophie Médiévale.* Louvain-la-Neuve: Institut Supérieur de Philosophie, 1988. 692–698.

Théodoridès, J. "Byzantine Science." In René Taton, ed., *History of Science: Ancient and Medieval Science from the Beginnings to 1450.* Translated by A. J. Pomerans. New York: Basic Books, 1963. 440–452.

Thijssen, J. M. M. Hans. "Some Reflections on Continuity and Transformation of Aristotelianism in Medieval (and Renaissance) Natural Philosophy." In *Documenti e Studi sulla tradizione filosofica medievale.* Centro Italiano di studi sull'alto medievo, Spoleto, II, 2 (1991): 503–528.

Thompson, James Westfall, and Edgar Nathaniel Johnson. *An Introduction to Medieval Europe, 300–1500.* New York: W. W. Norton & Co., 1937.

Thorndike, Lynn. *A History of Magic and Experimental Science.* 8 vols. New York: Columbia University Press, 1923–1958.

——— *University Records and Life in the Middle Ages.* New York: Columbia University Press, 1944.

——— ed. and trans. *The Sphere of Sacrobosco and Its Commentators.* Chicago: University of Chicago Press, 1949.

Wagner, David L., ed. *The Seven Liberal Arts in the Middle Ages.* Bloomington: Indiana University Press, 1983.

Wallace, William A. "Galileo and Scholastic Theories of Impetus." In A. Maierù and A. Paravicini Bagliani, eds., *Studi sul XIV secolo in memoria di Anneliese Maier.* Rome: Edizioni di Storia e Letteratura, 1981. 275–297.

——— *Prelude to Galileo: Essays on Medieval and Sixteenth-Century Sources of Galileo's Thought.* Dordrecht, Holland, and Boston: D. Reidel Publishing Co., 1981.

——— "Traditional Natural Philosophy." In Charles B. Schmitt and Quentin Skinner, eds., *The Cambridge History of Renaissance Philosophy.* Cambridge: Cambridge University Press, 1988. 201–235.

Wallace-Hadrill, D. S. *The Greek Patristic View of Nature*. Manchester: Manchester University Press, 1968.

Waterlow, Sarah. *Nature, Change, and Agency in Aristotle's "Physics": A Philosophical Study*. Oxford: Clarendon Press, 1982.

Watt, M. Montgomery. *The Faith and Practice of al-Ghazali*. Translated by M. Montgomery Watt. London: George Allen and Unwin, 1953.

*Islamic Philosophy and Theology*. Edinburgh: Edinburgh University Press, 1962.

Wéber, Edouard H. *La controverse de 1270 à l'Université de Paris et son retentissement sur la pensée de S. Thomas d'Aquin*. Bibliothèque thomiste, 40. Paris: Vrin, 1970.

Weinberg, Julius. *A Short History of Medieval Philosophy*. Princeton: Princeton University Press, 1965.

Weisheipl, James A. *The Development of Physical Theory in the Middle Ages*. London: Sheed and Ward, 1959.

"The Celestial Movers in Medieval Physics." *The Thomist* 24 (1961): 286–326.

"Curriculum of the Faculty of Arts at Oxford in the Early Fourteenth Century." *Mediaeval Studies* 26 (1964): 143–185.

"Developments in the Arts Curriculum at Oxford in the Early Fourteenth Century." *Mediaeval Studies* 28 (1966): 151–175.

"The Place of the Liberal Arts in the University Curriculum during the XIVth and XVth Centuries." In *Arts libéraux et philosophie au moyen âge. Actes du quatrième Congrès internationale de philosophie médiévale*. Montreal and Paris, 1969. 209–213.

"The Interpretation of Aristotle's Physics and the Science of Motion." In Kretzmann, Kenny, and Pinborg, *The Cambridge History of Later Medieval Philosophy*. 1982. 521–536.

*Nature and Motion in the Middle Ages*. William E. Carroll, ed. Studies in Philosophy and the History of Philosophy, 11. Washington, D.C.: Catholic University of America, 1985.

Wells, Colin M. *The Roman Empire*. Stanford, Calif.: Stanford University Press, 1984.

Wieruszowski, H. *The Medieval University: Masters, Students, Learning*. Princeton: Princeton University Press, 1966.

Wildberg, Christian. *Philoponus Against Aristotle on the Eternity of the World*. Ithaca, N.Y.: Cornell University Press, 1987.

*John Philoponus' Criticism of Aristotle's Theory of Aether*. Berlin: Walter de Gruyter, 1988.

Williams, John R. "The Cathedral Schools of Rheims in the 11th Century." *Speculum* 29 (1954): 661–677.

"The Cathedral School of Reims in the Time of Master Alberic, 1118–1136." *Traditio* 20 (1964): 93–114.

Wilson, G. N., ed. *Saint Basil on the Value of Greek Literature*. London: Duckworth, 1975.

Wippel, John F. "The Condemnation of 1270 and 1277 at Paris." *Journal of Medieval and Renaissance Studies* 7 (1977): 169–201.

"Did Thomas Aquinas Defend the Possibility of an Eternally Created World?" *Journal of the History of Philosophy* 19 (1981): 21–37.

"The Quodlibetal Question as a Distinctive Literary Genre." *Les Genres litté-*

*raires dans les sources théologiques et philosophiques médiévales. Définition, critique et exploitation. Actes du Colloque international de Louvain-la-Neuve 25–27 mai 1981.* Université Catholique de Louvain. Publications de l'institut d'études médiévales, 2ᵉ serie: textes, études, congrès. Louvain-la-Neuve, 1982. Vol. 5, 67–84.

Wissink, J. B. M, ed. *The Eternity of the World in the Thought of Thomas Aquinas and His Contemporaries.* Leiden: Brill, 1990.

Wolfson, Harry A. "The Problem of the Souls of the Spheres from the Byzantine Commentaries on Aristotle Through the Arabs and St. Thomas to Kepler." In Isadore Twersky and George H. Williams, eds., *Studies in the History of Philosophy and Religion.* Cambridge, Mass.: Harvard University Press, 1973. 22–59.

*The Philosophy of the Kalam.* Cambridge, Mass.: Harvard University Press, 1976.

Young, Charles R., ed. *The Twelfth-Century Renaissance.* New York, 1969.

Zaccagnini, G. "La vite dei maestri e degli scolari nello studio di Bologna nei secoli XIII e XIV." *Biblioteca dell' Archivum Romanicum,* ser. I, V. Geneva, 1926.

# Index

Page numbers cited directly after the semicolon following a final subentry refer to minor, unrelated mentions in the text of the main entry.